VODKA DISTILLED

Tony Abou-Ganim WITH Mary Elizabeth Faulkner

VODKA DISTILLED

THE MODERN MIXOLOGIST ON VODKA AND VODKA COCKTAILS

FOREWORD BY Dale DeGroff   PHOTOGRAPHY BY Tim Turner

To: Kristen
Happy Mixing!

S
SURREY BOOKS

AN AGATE IMPRINT

CHICAGO

Book design and layout: Brandtner Design

Printed in China.

Hardcover, first printing, January 2013

ISBN-13: 978-1-57284-125-3

ISBN-10: 1-57284-125-7

Library of Congress Cataloging-in-Publication Data is on file at the Library of Congress.

10 9 8 7 6 5 4 3 2 1

Surrey Books is an imprint of Agate Publishing. Agate books are available in bulk at discount prices. For more information, go to agatepublishing.com.

To Andrea and Carol—for moving the mountain to Mohammed at every turn.

Happiness!

Table of Contents

Foreword

VODKA IS ONE OF THE GREAT MARKETING STORIES OF THE 20TH CENTURY —a spirit as Russian as the communist revolution became the biggest-selling spirit in the United States at the height of the Cold War. Tony Abou-Ganim's thoughtful and comprehensive volume explores the rich tapestry behind the astonishing success of what is required by law to be an odorless and tasteless alcoholic beverage—but is it?

Tony and I traveled for over three years presenting Finishing School, a one-day vodka immersion course, speaking with bartenders about vodka and listening to what they had to say about the subject. We traveled all around the United States and to Canada, Mexico, Puerto Rico, even Australia, tasting vodkas and we answered the question posed above with a resounding "no!" Vodka is not a tasteless, odorless alcoholic beverage.

At each stop I was delighted by Tony's easy conversational delivery of the steps of vodka production, a subject that in lesser hands could easily replace counting sheep. In *Vodka Distilled* Tony captures that easy style in his prose and effortlessly gives us a clear picture of all the stages of production, styles by base ingredient, and—well, in essence he covers all the fine print in an appealing way.

Vodka remains the top-selling spirit in the United States, even in this era of big flavor when cocktails are characterized by the grafting of multiple culinary ingredients onto the root stock of the classic cocktail. Tony points out that vodka's primacy in the marketplace is beginning to get competition from a broad range of spirits embraced by the modern cocktail culture.

The past century encompassed two world wars and multiple regional conflicts, widespread movements to prohibit alcoholic beverages, the birth and death of the Soviet Union, body blows to capitalism beginning with the Great Depression and unprecedented climate upheavals around the world. Exploring new frontiers in the culinary world was low on the list of priorities.

And yet the last two decades of the 20th century were characterized by a relative measure of calm and prosperity in large parts of Asia, the United States, and Europe. The absence of widespread conflict allowed people to travel more freely and to explore culinary and lifestyle differences in other cultures.

Advances in technology connected ordinary people around the world and opened the door to experimentation in all culinary areas. In the United States a growing taste for big flavor fueled by this experimentation changed the US palate and led to an explosion of ethnic cuisines. The impact on the restaurant and grocery business, and eventually on the alcoholic beverage business, was significant.

While the Smirnoff brand dominated the category for over three decades, all would change in the mid-1970s when foreign imports like Finlandia, Stolichnaya, and Absolut created a demand for premium brands that would come to dominate the sale of vodka and eventually affect the wider category of spirit sales in general.

Tony deftly brings us up to the present moment and the unprecedented explosion in vodka production. Production costs in the vodka category are significantly lower than most spirit categories and that has opened the door for entrepreneurs who continue to grow the diversity and range of available products.

Most importantly Tony reminds us, as he has reminded the countless bartenders he has mentored, of the lesson he learned in Port Huron, Michigan, in a neighborhood bar operated by his cousin Helen from the end of prohibition until her passing in 2006: There are no short cuts to quality.

That lesson, cousin Helen, and his dedication to freshness in all the fruits, herbs, and savory ingredients used in his cocktails are the cornerstones of Tony's career and the focus of this book. Tony inspired a generation of bartenders and casino operators in Las Vegas with these same principles; with the beverage program at the Bellagio, Tony brought back quality cocktails in a town where cocktails were reduced to a free amenity and were delivered from a soda gun.

As Tony advises in the book,

> whether juicing; pressing; squeezing; muddling; garnishing;
> or making a syrup, purée, or infusion, don't even bother
> to arm wrestle Mother Nature. If the fruit is not at its peak
> and you wouldn't eat it fresh from the grocer's, don't mix
> it into your drinks . . . .

Finally, even with the subtle flavor nuances that the base ingredient, the water, and the distiller's art all bring to the spirit, vodka is a platform to deliver flavor, and quality is everything.

Tony has delivered an important book about vodka—a spirit that, believe it or not, has been instrumental in the rebirth of what author and historian David Wondrich calls the "first legitimate American culinary art": the cocktail.

Dale DeGroff

INTRODUCTION
For the Love of Vodka

THE MODERN MIXOLOGIST: CONTEMPORARY CLASSIC COCKTAILS **WAS A WILD** and wonderful journey. An exploration of my passion and professional underpinnings, and a colossal learning experience in how to follow one's dream of putting—rather pulling—together a book. Wouldn't trade it for anything—but, if I only knew what I was getting into!

Lesson number one—writing a book takes infinitely longer than expected. Lesson number two—as careful as we are to make it just right, every glance at the manuscript shows one thing or another that can be said a bit better. Lesson number three—typos and punctuation are slippery little things. Lesson number four—the last and by no means least—less is nearly always more.

Essential as it is to say something of interest, a stellar editor is as essential for keeping the message from drowning under the desire to cover every inch of a subject. So it was for us. End to end, Mary and I put a fearsome number of words to paper over the years as *The Modern Mixologist* evolved from one iteration to the next. In the end, sacrifices had to be made. Sobering surgery—but the result, we think, magic.

A chapter dedicated to the exploration of each spirit category was removed from *The Modern Mixologist*, which was indeed the right move. In our enthusiasm, we had included too much for the book's context but, at the same time, not enough to do the subject justice. Honoring each spirit—vodka, gin, tequila, rum, and whisky—deserves so much more than a page or two could offer.

If you have had the chance to enjoy it, you'll recognize *The Modern Mixologist* as a rich celebration of the cocktail experience, delivering how-to basics along with creative culinary inspiration to nourish your own bar-savvy talents. What better way to move forward than explore the nucleus of that experience? Celebrate each spirit at its singular best, where it lives within cocktail culture—and we hope your entertaining future. After all, where would the cocktail be except for the love of spirits?

WHY VODKA?

In part because numbers speak. Vodka is not only the number one consumed spirit in the United States but also ranks number one globally according to the Adult Beverage Resource Group at Technomic. And there is a certain compulsion to respond to the popular-yet-mistaken notion "Aren't all vodkas the same?" We've all heard it—if not uttered it ourselves before learning otherwise. That said, if I am being completely honest, the fact that vodka suffers from a misplaced lack of respect was highly motivating for me to write this book. Often passed over as a spirit category of interest, it is at times unjustly given a bad rap within the bartending community. Vodka's heritage and flavor nuances deserve a measure of reverence; it deserves a place alongside its spirituous cousins whisky, gin, tequila, and so on. Got to love a challenge!

Vodka Not Your Cup of Tea?

No hard feelings. In fact, nothing but respect for knowing your stance. That said, for reasons that largely escape me, it has become vogue in some bartending corners to bash vodka. Not just a particular brand or style, but the entire category. I assert such a stance most likely stems from lack of exposure or knowledge rather than a well-informed or experienced perspective.

We are all entitled to our preferences when it comes to enjoying sprits—flavor profiles, mixing compatibility in drinks, overall character. Such partiality evolves over time; it certainly has for me. But make no mistake, it has and always will be simply my personal preference, which admittedly is based on probably way too much exposure: 30-plus years of mixing, creating, teaching, and drinking in the bartending world. Exposure that I am fortunate enough to consider my professional experience and that thankfully continues to evolve.

Buck the trend. If you enjoy vodka to sip, or lean toward vodka-based cocktails and want to expand your appreciation of its unique position within mixology's sphere, I encourage you to ignore the voice that would persuade you to kick vodka to the curb as unworthy. I maintain such advice is less wisdom than simply a matter of opinion. Read on and enjoy!

Appreciating Subtlety

Despite the widely held view to the contrary, all vodkas are not the same. Qualities and variation from one to another can be subtle, but they are there nonetheless. Think about tasting and comparing one vodka to another, not as comparing apples to oranges but akin to comparing apples to apples—apples of the same variety grown in different orchards with differing geography and under various climate and nutrient conditions. All of these influences leave their subtle mark when the fruit is tasted. Wine geeks would call this the influence of *terroir*. So it is with vodka—subtle variations, but there nonetheless. Try a blind tasting, and you'll soon see this not-so-plain white spirit in a new light.

HERE AND NOW

Vodka Distilled is a celebration of vodka—for fans or fans to be—exploring the truly luscious and notable differences among various styles and brands.

Are you a Martini fan? Do you enjoy vodka neat or on the rocks with a twist? Is your drink of choice a big, bold Bloody Mary or a light, fruity poolside cooler? Are you a Cosmopolitan fan or partial to one of the stalwart classics such as the Gypsy Queen or Flame of Love? Whatever your tilt, each of these drinks can become an elevated experience with a fitting choice of vodka. Adhering to Modern Mixologist sensibilities, pay attention to every ingredient you put into your drink or cocktail. That means your selection of vodka as well. Bold and spicy or floral and light—it does make a difference!

Peeling Back the Onion

Where a spirit comes from—its heritage, institution, manufacturing process, and in particular its ingredients—is key to understanding what influences its quality, regardless and sometimes in spite of price.

This journey of peeling back the vodka onion, so to speak, starts with a bit of attention to vodka's story, including its history, manufacturing, and some production basics. Naturally, for your pleasure, I've included a selection of recipes—originals and classics—that could inspire you to create a few of your own. The concluding chapter is an examination of both the tasting process and the various vodkas available on the market, as a review of sorts. You'll see a profile of each brand's source, style, and character followed by tasting notes and recommendations for drinks from our panel of experts, charged with the task of blind tasting each and every vodka we selected—58 in all.

The robust selection of vodkas spans the spirit's full expressive range. You'll see both Old-World and New-World styles reflecting flavor profiles ranging from feminine, light, and approachable to masculine, spicy, and bold. All are categorized by ingredient base: rye, wheat, potato, mixed grain, and corn along with a collection of less traditional sources such as rice, grapes, and whey, to name a few. They are categorized this way because this is a spirit composed essentially of two ingredients: water and a fermentable sugar source. It is arguably this ingredient base, or fermentable sugar source, that carries the most weight in how vodka expresses itself. The combination of water, fermentation, distillation, rectification, and filtration rounds out its character, taste, and mouth feel.

To be clear, our intent is in no way an attempt to identify which vodka is best or to rank them top to bottom. By offering each brand's flavor profile, qualities, and strengths, we simply hope to illustrate how to identify what vodka, or style thereof, is best for you. Packaging and pedigree aside, vodka is something only you can judge. In the end, vodka selection depends entirely on individual taste and preference.

Bottom Line

While this brief guide may not elevate you to expert status, it will we hope deliver ample information and inspiration to turn you at the very least into a better-informed fan, motivated to explore and create. *Vodka Distilled* is meant to serve as a resource and inspiration for finding the perfect selection to star in your favorite vodka-centric drinks or for discovering a new gem you will simply and happily pull from the freezer to sip straight from a frosty chilled glass.

Get ready to host your own tasting and to try a few recipes on for size. We hope they'll become some of your new favorites or favorites revisited. Beauty, after all, is in the eye of the beholder—or in this case, on the beholder's palate!

CHAPTER ONE

The History of Vodka

THE DISTINCT RIVALS FOR THE CROWNING TITLE "VODKA'S BIRTHPLACE" are definitely Russia and Poland. William Pokhlebkin, in his extremely comprehensive book *A History of Vodka,* offers a thorough and compelling case for his native Russia. He argues that historical documentation establishes vodka production in that country as early as the 14th century, while his research placed Poland's first vodka production considerably later—in the early to mid-16th century. As exhaustive as his research is, Pokhlebkin's assertions are not entirely unopposed. To a certain degree the question of who was first is moot. In 1982 the Soviet Union was declared the first to have produced vodka—by an international tribunal. Of course it is not that easily resolved for many, and the debate continues.

The principal problem with taking a firm stand on who was in fact first to produce vodka is that many historical references are conflicting. Further, in the case of many European countries, including both Poland and Russia, the existence of alcoholic drinks or tinctures, known by one name or another, is well documented, dating back much further than the 14th century. The word "vodka" itself—or "voda" in Russian and "woda" in Polish—means water and offers little insight as to when or where the grain-based distillate first appeared. A variety of sources indicate that some of the earliest vodkas or grain-based wines were referenced in historical documents as early as the ninth century. The Vyatka Chronicle of 1174, for example, is said to document the existence of a vodka distillery at Khylnovsk, Russia—some centuries earlier than even Pokhlebkin's findings.

Here is where the waters become further muddied because of confusing terminology. Using the term "distillery" is problematic because it implies people in the

12th century were using the same process of distillation that we use today, and that they were producing a grain-based spirit and enjoying it as a beverage. Further exploration indicates, though, that this seems not to have been the case until some three centuries later. It is better to consider the term "distillation" in 1174 as being synonymous with alcohol production, which would have occurred by rudimentary 12th century means and for more utilitarian purposes. In any case, in the centuries leading up to the 14th, the stage had been set—vodka's heredity was incubating.

EARLIEST INFLUENCES

Evidence indicates early distillates are tied to monastic uses of *aqua vitae* or "water of life." As Desmond Begg explains in *The Vodka Companion: A Connoisseur's Guide*, "monks [were the ones] who kept the secrets of distillation alive" through the dark ages; Begg gives them credit for reviving distillation on the continent "after AD 1000." The earliest spirits were more closely related to an eau-de-vie, or grape-based distilled wine, rather than a grain-based potion, as vodka would eventually be defined by some. Versatile and widely used, these vodka precursors were less beverage and more utilitarian product. Aside from their medicinal, monastic, and industrial applications, both internal and topical, early spirits were even used as cosmetics—including as a form of aftershave according to Dave Broom in *The Connoisseur's Book of Spirits & Cocktails*.

Hairsplitting will no doubt continue, but the point at which earlier versions of vodka transitioned from being utilitarian to being an enjoyable beverage was close to the 15th century. At this point distillation techniques became more refined. Methods used in Italian monasteries, specifically pot distillation of wine, were studied by visiting Russian monks. A delegation is known to have journeyed to Italy in 1430. No doubt the relatively refined Italian distillate inspired the Russians' application of similar techniques when they returned home. Broom points out that the delegation's return "coincided with a grain surplus," reinforcing the mid-15th century timeline for the establishment of Russian vodka distillation.

As distillation techniques were practiced and eventually refined, vodka's potential for revenue as a consumable was captured. As Stuart Walton asserts in *Vodka Classified: A Vodka Lover's Companion*, "In due course distillation spread from the monasteries outwards into the homes of peasant families." By the 1470s Tsar Ivan III had imposed the first state control over distillation—and by association its revenue. A grasp the state would firmly hold until the 1700s. Similarly, and during the same time frame, Poland's nobility was granted sole rights for vodka production within its territories—along with the ensuing considerable financial prospects.

Did vodka exist prior to the 1400s in Russia and Poland? In one rudimentary form or another, this argument is indeed compelling. But it's a question best left to the experts. For our purposes, better to look forward. Where did vodka voyage hence, what sources influenced it, and what influences did it exert along the way?

EVOLUTION, PAST AND PRESENT

Vodka's position as a consumable for pleasure is undeniably rooted in the Vodka Belt—a modern-day reference to the collection of European countries or regions known for their heritage and tradition as vodka producers. By majority consensus the Belt includes Russia; Poland; Belarus; Ukraine; the Nordic states of Sweden, Norway, Finland, Iceland, and Greenland; the Baltic states of Estonia, Latvia, and Lithuania; and parts of Slovakia and Hungary. I suspect debate exists here as well—about whom to include or exclude.

The history of how vodka production emerged within each state or region is difficult to trace with precision. The intricate histories traced in Russia and Poland turn out not to be unique, as others within the Vodka Belt share a similar experience. Each region has its native tradition surrounding alcohol production, use, and consumption—be it medicinal, monastic, or some other application. The building blocks were clearly present for each culture. Imagine these early regional customs being nurtured over time by an array of influences—political, cultural, agricultural, and mechanical—and each one's distillation heritage developing in its own way, at its own pace. Vodka's story represents a path that is neither strictly linear nor chronological, rather one of a thousand influences combined over time, place, and circumstance.

First Steps: Russia and Poland

Vodka proper, manufactured as a consumable for profit, appears to have migrated from Russia in the 16th century. By this time, Begg notes that "Moscow was producing enough [vodka] to be able to export to Sweden and Estonia." The state realized that vodka was an established consumable with lucrative prospects. Access and consumption increased, though quality did not necessarily improve for the commoner. The masses drank vodka made of any accessible agricultural resource, including beets, potatoes, and even nettles. By contrast, the nobility, flush with wealth and means, enjoyed a premium product typically made of rye. Both arms of this consumer dichotomy inextricably secured vodka's place within the Russian ethos over the centuries and its eventual journey beyond Russia's borders.

Aptly characterized by Linda Himelstein in *The King of Vodka: The Story of Pyotr Smirnov and the Upheaval of an Empire*, vodka became "as fundamental to daily life as food and the wintry chill." Culturally embedded both deep and wide, it held a position in all corners of Russian life. Serving as an emblem with many facets— success, reward, incentive, remuneration, sustenance, celebration and ritual— vodka was more than just a national beverage, it was an "entrenched national habit." A reality the infamous and fiercely entrepreneurial Pyotr Smirnov would successfully harness. Engineering his brand into the largest vodka producer in the country and himself into one of the wealthiest men in Russia, Smirnov reached far beyond what even his voracious appetite for achievement could imagine.

Tsarina Catherine II, also known as Catherine the Great, is another colorful character who played a significant role in extending the reach of her beloved spirit beyond Russian lands. As Walton explains, Catherine, who reigned in the 18th century, was a known "connoisseur of vodka," active and industrious in her promotion "of Russian culture." She was also known for her "distinctly Western-oriented intellectual tastes," and her letters show that she corresponded with and enjoyed the company of many great intellectuals of her era. Her influence touched prominent figures residing in Europe's other leading countries, which effectively propagated an appreciation for the spirit's more refined expression.

Poland's culture was similarly infused. Vodka's leap from utilitarian liquid to revenue-producing potable occurred primarily in the 16th century, with exports firmly established over the next century. By this time vodka was widely available and consumed for pleasure, and its manufacture had made the transition from cottage industry to a commercial one. Begg explains that the scale of vodka production transformed in the middle 1500s when vodka produced plentifully in Krakow was exported to neighboring crown lands. Some 30 years later the epicenter of Poland's production would move northwest to Poznan, which eventually housed numerous distilleries (and remains a major center for vodka production today). From here the industry flourished. By the end of the 16th century, people were drinking Polish vodka not only in neighboring crown lands but also throughout much of Europe and Russia.

As an industry, Poland's vodka production continued to evolve. Becoming well versed in the craft of distilling both neat and flavored products, others took notice. Thanks to ongoing ingenuity during the 18th century that led to increasingly refined distillation and filtration techniques, "Polish vodka became the model for quality production across Eastern Europe, with equipment (and techniques) being exported to Russia and Sweden," shares Broom.

Over time, as with most of the Baltic and European regions, wars, strained economic realities, and political relationships all helped to navigate vodka's journey—for better or worse. Yet its eminence never wavered within Poland's cultural margins. Consumption remained steady through thin and thick. Shortages in the 1980s merely influenced a notable trend in home distilling. Within a decade, privatization revived the industry to national dominance and considerable international reach. Specialty and flavored vodkas included, Poland currently produces over 1,000 different brands according to Broom, and the country's national consumption rate is among the highest in the world.

Sweden, Finland, and Beyond

Brännvin, which translates to "burnt wine," is the vodka predecessor—rather equivalent—native to Sweden. Similar to vodka produced in other European regions, it was a spirit distilled from potato, grain and in its earliest form probably even from wood byproducts. (Sweden's brännvin would not be known by the name "vodka" until the 1950s). Dating back to the 15th century, it too was primarily produced for medicinal use and it was a key component in the manufacture of gunpowder; the latter use led to strict regulations of the spirit. As Begg asserts, after brännvin's intoxicating qualities were realized by the public, overconsumption began to adversely influence supplies needed to produce gunpowder. In response, Stockholm's city council threatened to seize both equipment and product from anyone selling their own spirit without a license.

Brännvin continued to gain popularity, but its move from a beverage affordable for a select minority to one accessible to commoners did not occur until much later—in the 16th century. By the 17th century brännvin reached the masses as "the people's drink"—so dubbed by Nicholas Faith and Ian Wisniewski in *Classic Vodka*.

In the centuries to come, Sweden's brännvin (vodka) production and drinking culture evolved drastically. What became Sweden's most familiar vodka export—known today as Absolut Vodka—was launched in 1877 by Lars Olsson Smith, a larger-than-life character who, according to Walton, "bestrode the works of Scandinavian spirit production like a colossus." Driven and obsessive, he is most notable for having developed a spirit with as yet unknown purity. He called it Absolut Rent Brännvin, or "absolutely pure vodka," which Broom calls "Sweden's first rectified spirit." Under Smith's persuasive leadership the brand prospered, but strict state regulations halted the spread of a commercial empire.

Sweden's vodka industry suffered greatly during the early to mid 20th century. Prey to the strains of two world wars, they also weathered the choking grasp of a temperance movement roused by high rates of alcohol consumption. This cultural development was adopted by the state, which took complete control after World

War I by imposing impressive taxes on alcohol producers and closely rationing consumers. The V&S was established (Vin and Sprit or Swedish Wine and Spirit corporation) to regulate all aspects of the alcohol industry.

The group aimed to curtail overconsumption and continued rationing into the 1950s. Its success was largely tempered by the fact that home distillation thrived and trips across any border where drinking was without restriction were commonplace. In 1995 Sweden became a member of the European Union (EU) and vodka production was released from the state's monopoly – though control over retailing remained within its domain. Ultimately, Absolut was resurrected with great success in 1979 under the V&S umbrella organization and has continued to thrive. Considered exemplary, particularly for its marketing prowess, it remains the flagship of Sweden's vodka exports and is among the world's most successful brands.

Finland is another Vodka Belt powerhouse. According to Broom, Finnish vodka was produced as early as the 16th century. This notion is supported by Begg, who speculates the Finns most likely learned the craft of vodka distillation through "mercenaries returning home from European wars" during the time. With grain and water both in abundance, vodka production quickly caught on. As Faith and Wisniewski note, even though vodka was "usually produced in home stills . . . by end of the 17th century, it had almost replaced beer as the national drink."

It was this exact enthusiasm that led to a shortage of yeast, crippling vodka production. Eventually supplies could not meet the demands of consumption. This crisis had implications that continue to shape Finland's vodka industry today. The shortage led in short order to the creation of a factory in Rajamäki in the 1880s dedicated to yeast production. This factory became the eventual partner of the country's only surviving vodka distillery.

Fast-forward through several wars, throw in struggles with a growing temperance movement parallel to that in Sweden, and the industry dramatically evolved. Production became more centralized, moving to the northwest of the country where water and grain were more readily available. Here the Koskenkorva distillery was built, and it remains Finland's only extant vodka-producing facility. They are best known internationally as the producers of Finlandia, which is distilled in Koskenkorva and bottled in the aforementioned factory in Rajamäki. Launched in 1970 specifically for the export market, Finlandia is recognized and readily available worldwide.

Newcomers

Denmark, Germany, Switzerland, France, and the Netherlands—each one has its long, rich history of distilling spirits, white and dark. Vodka proper, however, is a relatively recent phenomenon, not appearing for the most part until the 20th century. Though comparatively short on legacy, vodkas produced by these other Western European nations are not inferior. On the contrary, I agree with Begg, who concludes that the distillation and rectification techniques refined over centuries for distillates—such as calvados, jenever, akvavit, kirsch, cognac, armagnac, whisky, and other spirits traditional to these regions—represent established and accomplished practices that translate beautifully when applied to vodka production. In this modern paradigm, you can safely discard pedigree; Ketel One, a product of the Netherlands, is a prime example. In 1983 the Nolet family employed spirit distillation practices dating back some 300 years to develop what is widely considered the world's first ultra-premium vodka—a concept the industry quickly and passionately embraced.

THE UNITED STATES DISCOVERS VODKA

Despite its prominence today, vodka is a relative newcomer to the mainstream drinks culture in the United States. The assumption is widely held that Smirnoff, in 1934, was the first vodka import to these shores. Alas, another preceded it: "Wolfschmidt distilleries had tried decades earlier and been met with universal disinterest," explain Susan Waggoner and Robert Markel in *Make Mine Vodka: 250 Classic Cocktails and Cutting-Edge Infusions*. The initial attempt to break into the new market was likely targeted at immigrants from Eastern Europe, but Wolfschmidt regrettably found an opportunity whose time had yet to come. Even the Smirnoff company found itself up against imposing odds upon its US launch and subsequently struggled to gain traction among consumers—this was the start of what became an extensively documented test of perseverance and providence, and with extraordinary consequences.

The story of vodka's journey to the United States truly begins much earlier, set in motion in large part by the rise of Leninist communism. In Russia in 1917, the Bolshevik government confiscated and subsequently closed the extraordinarily successful Smirnov distillery. Himelstein explains that Vladimir Smirnov, the third son of distillery founder Pyotr Smirnov and who was in charge of the family business at the time, was declared "an enemy of the people." A despised capitalist, he tussled futilely before the unyielding tidal forces of communism. His wealth represented an enticing target for the Leninists, particularly because it was tied to vodka; the Leninists blamed the consumption of vodka for corrupting the Russian people. The Bolsheviks arrested Smirnov, took him prisoner, and quickly sentenced him to death.

Though circumstances are unclear, Smirnov escaped his captivity and fled to Turkey, where he set up a distillery in Constantinople. Production moved to the city of Lviv—then a part of Poland but currently in Ukraine—and a second distillery opened in Paris. At this point in the 1920s, the brand changed from its native Smirnov to the now more familiar transliteration Smirnoff.

Selling vodka to the French proved challenging. Anistatia Miller and Jared Brown explain in *Spirituous Journey: A History of Drink* that Smirnov failed miserably at convincing the locals to drink a "tasteless, colorless, ethnic spirit such as vodka." In part the French were repulsed by a spirit reminiscent of communist politics. In 1933 Smirnov sold the name, recipe, and rights to produce Smirnoff in the United States to Rudolph Kunett, a Ukranian-American visitor to Paris who befriended Smirnov with ease. Kunett was keenly familiar with its potential, as his family supplied spirits to Smirnov in pre-revolution Moscow.

The United States was struggling to emerge from its prohibition years, yet Kunett saw an opportunity. In 1934 he moved production to the United States, setting up a distillery in Bethel, Connecticut. This was a purposeful move, as Begg points out, since Bethel was "an area with sizeable Russian and Polish communities." This move alone may have been the most important factor in what sustained him at all.

The critical flaw in Kunett's plan was not realizing US consumers weren't in the least bit familiar with the virtues of vodka. They preferred instead another white spirit to which their palate was already accustomed: gin. Gin was readily available at the close of prohibition and drinkers weren't looking for anything different. This established taste was a barrier that proved difficult to remove. This chapter in the Smirnoff journey ended in 1939 when Kunett, some five years after bringing the company to US soil, sold the business to John G. Martin, President of Heublein Inc., for $14,000—a paltry sum in retrospect.

Still struggling, production moved to Hartford, Connecticut, in 1939. Here the floundering brand enjoyed a boost, but it may have been a fluke rather than the relocation that provided the spark. The company used mislabeled corks marked Smirnoff Whiskey and marketed the product as a whiskey "without flavor, odor or color." This concept began to resonate with the US drinker. But as Waggoner and Markel remind us, this would be a brief rise in popularity, because the company's growing success ran up against World War II when "distillers converted their production lines to war work, and liquor was a rare commodity."

Eventually vodka's value and versatility as a spirit that could be mixed with just about anything was realized. This transition came about because of the friendship and collaboration between John G. Martin, Heublein's president, and Jack Morgan, owner of the Hollywood restaurant the Cock 'n Bull. Morgan had endeavored to import and introduce British ginger beer to the US market but found it a very hard sell; the locals weren't interested. Martin, similarly, had his floundering product to promote. Fueled by mutual commiseration and determination, they decided to combine the two—ginger beer and vodka—add in some lime juice, and market an entirely new offering. They called it the Moscow Mule. Morgan's girlfriend rounded out the drink's concept: According to Ted Haigh in *Vintage Spirits & Forgotten Cocktails*, her company produced copper products and created the namesake copper mug embellished with a kicking mule.

A combination of happenstance and ingenuity, the genesis of this one drink was a pivotal development nonetheless. According to William Grimes in *Straight Up or On the Rocks*, "For the first time, an invented cocktail was being used as a marketing device." This was a key step toward capitalizing on the relationship between consumption and marketing, which in time propelled the industry to unimaginable heights worldwide.

Heublein's marketing efforts to promote the Moscow Mule proved successful, finally placing vodka into the US drinking lexicon. Its growth continued during the 1940s and made exponential strides in the 1950s. Grimes estimates that between 1950 and 1954, sales leapt from 40,000 cases to over 1 million and then more than tripled the following year.

From that point forward vodka—as a spirit category and a mainstay in the US drinks culture—has never looked back. Today the vodka industry as a whole has surpassed the $12 billion mark annually.

CHAPTER TWO

Vodka Defined

IT IS STRIKING HOW FAR REMOVED THE US DEFINITION IS FROM WHAT vodka strives to be in its ancestral countries. For the Vodka Belt, the concept of flavor neutrality is largely misplaced, rooted instead in a legacy that ultimately celebrates flavor instead. Given the context in which these different vodka cultures grew—Old World versus New World—such a vast dichotomy in definition is not surprising. It is in many ways rational, though.

EUROPEAN UNION

The EU currently defines vodka in decidedly straightforward—even simplistic—terms. It must contain a minimum of 37.5 percent alcohol by volume and is otherwise "a spirit drink produced from ethyl alcohol of agricultural origin." Who would have thought that one sentence would become so provocative?

Prior to joining the EU, the Vodka Belt countries—led primarily by Poland, Sweden, and Finland—voiced concern over this definition, arguing it did not protect traditional vodka producers' interests. Specifically it allowed room for vodka to be made from fermentable sugar sources other than the more traditional sources of potato, grain, and the lesser-advertised molasses, which manufacturers had long used in the production of sub-premium vodkas. These were understandable concerns—vodka's long-standing traditions were being threatened. Conversely so was the industry itself, a formidable entity that is worth billions.

Prior to entering the EU in 1995 the Vodka Belt contingent requested the European Parliament to adopt a tighter definition of what may be designated "vodka"—a

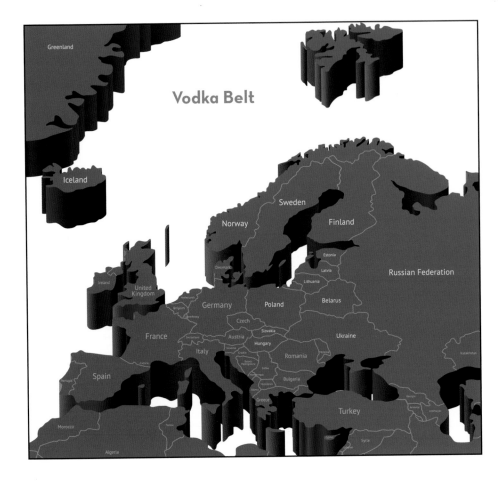

Vodka Belt

step that guards the interests of *true* vodka producers and protects consumers from being misled. Vodka, the contingent argued, should be given the same protection offered other spirits such as tequila, whisky, and brandy. Specifically, the producers sought the "designation of origin" protection. Ingredients and manufacture processes for such distillates are clearly defined, sheltering them from competition with spirits made of inferior or cheaper ingredients and methods. A familiar regional example is champagne, which can only be produced within clearly defined geographic boundaries. Any product made outside that region may neither wear nor profit from the designation "champagne."

From here the Vodka Wars ensued with Vodka Belt countries on one side and the EU and many other Western nations on the other. Traditional producers maintained that the designation "vodka" should only be given to products made in traditional Vodka Belt countries, and only those made with traditional ingredients—potato, grain, and molasses. The opposition including Britain, the Netherlands, France, Italy, and Spain among others fiercely opposed this move, because it would force established brands from outside the Belt to either dissolve or reconfigure as a different spirit altogether. The financial repercussions would be vast.

Enter the Schnellhardt compromise. Horst Schnnellhardt, a member of the European Parliament for Germany, proposed that vodkas made of anything other than potato, grain, and molasses be labeled "Vodka produced from . . . " followed by the specific base ingredient. This was far from satisfactory to the Vodka Belt contingent, which counter-proposed that all nontraditional brands be required to clarify their nontraditional status with wording of a size that covered the majority of the product label. The European Parliament would not accede. In 2007 it adopted the Schnellhardt compromise but did not change the incumbent definition of vodka, which remains "a spirit drink produced from ethyl alcohol of agricultural origin." Hailed by the Western opposition as a success, this outcome left the Vodka Belt countries feeling the sting of defeat. Efforts to pursue their traditionalist interests continue today.

RUSSIA

As a Vodka Belt behemoth operating well outside EU restrictions, Russia has no interest in reforming its long established definition of vodka. The country's definition is founded on the research of 18th century scientist Dmitri Mendeleev, who was named director of the Imperial Bureau of Weights and Measures in 1893. Known for his interest in securing the perfect concentration for vodka— that is the proportion of water to alcohol, or proof—Mendeleev discovered precision was more difficult to attain when measuring by volume; measures of weight proved considerably more reliable. In the end he established the ideal proportion as 38 percent alcohol to 62 percent water. These percentages were subsequently rounded off to 40 percent and 60 percent, respectively, to simplify taxation. According to Pokhlebkin, from this point on "Russian vodka—or, more precisely, Moscow vodka—came to be defined as a grain spirit, triple distilled and then diluted with water to a concentration of 40 percent by weight." Established officially by the Russian government in 1894, it remains the national standard.

UNITED STATES

The US definition of vodka offers an entirely different perspective. Vodka is described in Subpart C, Section 5.22, Item 1 of the *Standards of Identity for Distilled Spirits* as "A neutral spirit, so distilled, or so treated after distillation with charcoal and other materials as to be without distinctive character, aroma, taste or color." It's not difficult to see how this definition mirrors vodka's development in the US spirits market.

Vodka's value in the United States is largely based on the idea that its taste is less detectable than that of other spirits and it has a huge capacity for mixability. Principally influenced by the arrival of Smirnoff on the scene, these two concepts combined to inspire a trend toward mixed drinks and cocktail consumption that spawned an industry of titanic proportions. Smirnoff's "it leaves you breathless" add campaign in the 1950s, featuring the Moscow Mule and espousing vodka as a spirit with neutral flavor, not only embraced but in large part influenced these Western sensibilities right from the beginning. In contrast, traditional, Old-World production and consumption patterns reveal that vodka served a much broader purpose in the Vodka Belt.

OLD WORLD VERSUS NEW WORLD

The vodka category has two opposite ends of flavor and character. That is to say, vodkas may be characterized as either traditional or modern, Eastern European or Western European (and beyond), masculine or feminine, or Old World or New World. But there is a certain amount of crossover along this spectrum, depending on what aspect of the vodka is being considered, such as production technique, geographic origin, and character and flavor profile.

Without a bit of instruction on how to decipher these interchangeable terms, evaluating the broad collection of vodka brands becomes a jumble fast. In the interest of minimizing confusion, I'll share my interpretation, which those with more knowledge may feel free to challenge.

Traditional, Eastern European, or Old-World Vodka

Old World refers to vodka produced in the Vodka Belt. As noted in Chapter One, the Belt consists of European countries or regions known for their heritage and tradition of vodka production, distillation method, and consumption practice—all of which cultivated a style that veers toward the flavorful end of the spectrum.

In these regions, vodka distillation was largely developed from the early practice of using locally available materials—rye, wheat, barley, and potato—to start. The goal was *not* to produce a flavorless distillate, merely one with bitter or undesirable elements removed. Hence, unlike the modern, Western, or new-world style, old-world vodkas are more likely to retain the characteristics of their base ingredient. They are bold, flavor forward, even robust or spicy.

It makes perfect sense. For centuries the practice has been to enjoy and savor a glass of vodka neat "with the same respect one might lavish on fine brandy or single malt Scotch Whisky," as Anthony Dias Blue depicts in *The Complete Book of Mixed Drinks.* Hence, for this practice, vodka that has more depth or more prominent flavor and character profile is desirable. This is what I consider Old World in style and character. In the interest of full disclosure, as the industry has evolved, some producers from within the Vodka Belt have turned out vodka that is in tune with the Western palate, which is less inclined to look for fire and spice over neutrality.

Modern, Western European, or New-World Vodka

Odorless, colorless, and tasteless style of vodka is a modern, Western, or new-world concept. I consider vodkas New World by virtue of their neutral flavors and not by their modern distillation techniques, use of nontraditional base ingredient, or even Western geographic genealogy. This flavor neutrality developed initially in the West within the past century and is largely reinforced by the unique relatively adolescent drinks culture that has evolved in the United States since the late 1940s and early 1950s.

I characterize New-World vodka as being more approachable, neutral to light in character, and tending toward the feminine end of the flavor spectrum. Here, again, culture largely dictates practice. What caught on in the West, specifically the United States, was a white spirit boasting such a neutrality of character that it could be mixed with anything and remain largely undetectable.

It All Comes Back to Personal Preference

With this general explanation, you can begin to get a feel for how vodka style and character take form as a solid basis for comparing different vodkas, and developing your preference. Is Old World better than New World? The answer to that is entirely the consumer's decision, based on what best suits their palate and drinking practices. The Old World/New World concepts are merely guides, posted at opposite ends of the spectrum. Again, some vodkas produced in the traditional Vodka Belt veer toward the lighter, New-World style. In contrast, some vodkas made in modern, Western regions intentionally have Old-World style, spice, and character.

Drawing out the distinction between Old World and New World adds layers of complexity that is beyond the scope of our discussion. For our purposes moving forward it is simpler to establish Old World and New World in terms of flavor profile and overall character, rather than in terms of geographic origin, distillation method, or pedigree of a particular brand or distillery. Being clear on this basic difference is particularly important to fully appreciate the tasting notes in Chapter Five, where our tasting panel explores each brand's distinctive style and character rather than ranking them from top to bottom.

CHAPTER THREE

The Anatomy of Vodka

VODKA IS AMONG THE LEAST COMPLICATED DISTILLATES TO MANUFACTURE. Its primary influences are the raw materials used (water, base ingredient or fermentable sugar source), method of fermentation, distillation and rectification, process of filtration, and water used as diluent.

Compared to manufacturing dark spirits such as whisky, tequila, and brandy, vodka production stands out in two distinct ways. First, the distillation process aggressively eliminates elements that influence flavor (often called "congeners"), which can be desirable in other spirits. Second, the final product is free of the considerable influences of maturation or barrel aging. To a large degree, this makes vodka a more exposed, less forgiving distillate, as impurities and imperfections that occur in the manufacturing process are more detectable in vodka than in spirits influenced by the complexities of aging.

This book aims to highlight vodka's virtues, so my discussion of the production process is decidedly basic, exposing points in the process that exert discernable influences on the final product. For an in-depth explanation of spirit manufacturing, check out the countless resources dedicated to the craft, whatever your spirit of choice.

RAW MATERIALS

The base ingredients used initially were those most readily and locally available, though that practice evolved as the industry sought more innovative approaches. Vodka can be made from virtually any starch, or fermentable sugar source.

Traditionally, as remains the case in Eastern Europe, grain—such as rye, wheat, or barley—and potato is used. The rest of the world, however, incorporates what traditionalists call "alternative" ingredients, such as corn or oats. Grapes, whey, maple sap, rice, and quinoa are among the less familiar and more recent additions. One of the least expensive and little advertised base ingredient is molasses. Derived from sugar beets and sugarcane, it has long been used in the production of large-volume brands that target the bulk or sub-premium market —common practice in both the Vodka Belt and the West.

Bottom line: A quality distillate starts with quality ingredients. The base ingredients selected impart a constellation of characteristics to the final product. In general, though by no means the rule, the four most common base ingredient categories—rye, wheat, potato, and corn—fall along lines of expression for nose, palate, and mouth feel distinct to each (see page 130 for the typical vodka-tasting findings related to nose, palate and mouth feel according to base ingredient).

Water is arguably the second raw material used in vodka production. I discuss its use and influence toward the final product further along in the Dilution section.

FERMENTATION

The first step in making any distilled spirit is to produce the equivalent of a fermented alcoholic beverage called a "wash." This is achieved by converting the raw material's starch to fermentable sugars. Fermentation is initiated by adding malted grains into a "mash" of water and raw material. The naturally occurring enzymes—alpha amylase and beta amylase—in malted grains serve as a catalyst for the conversion process, brought along by adding heat. The mash now contains simple fermentable sugars—glucose, fructose, and sucrose. Yeast is then added, which essentially feeds on the sugars. The quality and consistency of the yeast used can influence the quality of the vodka produced, so care must be taken when selecting yeast to ensure the best possible final result. The metabolite byproducts of the fermentation process are ethanol (alcohol) and carbon dioxide.

It is this fermented mash, now referred to as wash, that proceeds to distillation. Made up of a relatively low percentage of alcohol by volume (ABV), water, and a small component of impurities, the wash is similar to a beer product—raw, unrefined, but full of potential. How this potential is realized relies largely on the distillation and rectification skills of the Master Distiller.

DISTILLATION AND RECTIFICATION

In the most basic of terms, distillation concentrates alcohol to a desired level, while rectification serves to purify it.

The distillation process separates components of a liquid mixture by using boiling points. In the case of vodka, the wash's alcohol (ethanol) component is separated from its water component. Alcohol has a lower boiling point than water—meaning, it vaporizes at a lower temperature—so alcohol can be collected separately from the water.

Rectification, meanwhile, refers to the process of purifying alcohol of its unwanted elements. Although the term is more commonly associated with continuous distillation than with pot or batch distillation (read on for this discussion), rectification is, in essence, achieved through repeated distillations, leaving the concentrated alcohol free of waste and unwanted elements—such as congeners (impurities), fusel oils, methanol, and aldehydes that would adversely affect the quality—flavor, aroma and texture—of the final product.

The two basic methods of spirits distillation—pot (batch) and column (continuous)—achieve the same ends and require the same considerable expertise of a Master Distiller, though continuous is arguably more efficient. In contrast to distillation of dark spirits—whisky, brandy, tequila, and so on—vodka distillation aims to get rid of nearly all congeners and leave only the trace amounts needed to express a desirable level of flavor, aroma, and texture. Of note, it is these removed elements that contribute largely to the flavor profile of dark spirits—and likely to the depth of any hangover.

Pot or Batch Distillation

Before the column or Coffey still was invented in the 19th century, distillation was done in a pot still. The pot still, sometimes called an "alembic," was used as early as the third century in Alexandria. This still has two separate chambers—pot and condenser—connected by a tube or swan's neck. Fermented mash, at roughly 8 percent ABV, is poured into the belly of the pot still and gradually heated. Once the temperature of the mash reaches the alcohol's boiling point, vapors rise upward to the swan's neck where the vapors cool. The vapors then begin to travel down the swan's neck into the condenser, where the process of returning the alcohol vapors into liquid form is completed.

The first portion or fraction of pot distillation is called the "heads" and is extremely high in impurities, such as methanol, aldehydes, and esters. The Master Distiller cuts (removes) this portion of the distillate and discards it. The middle portion is referred to as the "heart" and is desirable; the heart is what the Master Distiller works to maintain as the final product. The last portion is called the "tails" and contains unwanted fusel oils and is high in water content; the tails must also be cut and discarded. A single run yields a distillate of approximately 40 percent ABV.

From this point forward the Master Distiller and his proprietary domain rule. The whole process can be repeated many times—by reintroducing the condensate or heart back into the pot for another run—or not at all, as deemed desirable for revealing the heart of ideal quality. Again, ideal is a matter of preference. Fewer runs mean more flavor and character are left behind. More runs yield a distillate of greater purity, flavor neutrality, and increasing alcohol content that could be as much as 96 percent ABV in the end. Rarely consumed at such a strength, the concentrated distillate has to be cut with water, usually prior to the filtration process, to dilute it down to the acceptable minimum bottle strength for vodka—between 37.5 percent ABV (in the European Union) and 40 percent ABV (in the United States).

Pot Still

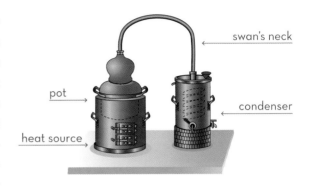

Pot distillation is usually more labor intensive than column or continuous distillation, particularly if a batch is redistilled multiple times. I argue that where the Master Distiller cuts the heads and the tails is more important than the number of distillation runs. Achieving quality spirit is the main goal and depends on how each producer realizes its desired balance of purity and flavor.

Column or Continuous Distillation

Invented by Robert Stein in the 1820s, the column still was first used in Scotch whisky production. In 1831 Aeneas Coffey earned a patent for his improved version, known as the Coffey still or patent still. Regardless of name, the column or continuous still is made up of two cylinders—the analyzer and the rectifier—placed side by side and connected by multiple pipes. In this system the mash is fed into the still, proceeds through the distillation and rectification process, and transforms into a distillate of desired strength and purity—all in a constant and seamless motion. This efficiency is unparalleled. You could manufacture a

Column Still

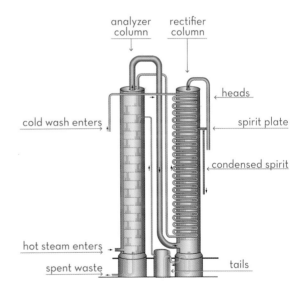

greater amount of spirits at a faster speed with column distillation than with a series of sequential pot distillations.

Describing the process of column distillation simply and without being too technical is a challenge. Begg relays this scientific process in straightforward terms. I've summarized the basic steps:

Cold wash enters the top of the rectifier column, traveling along a closed pipe system located within the column, where it becomes heated before being pumped into the top of the analyzer column.

Hot steam enters the bottom of the analyzer column, which is separated into horizontal sections by a series of plates.

As the steam rises within the column, the horizontal plates become heated.

Heated wash is pumped into the top of the analyzer, coming into contact with the steam-heated horizontal plates on the way down.

Once the wash makes contact with the first set of heated plates, alcohol in the wash begins to vaporize.

Desirable alcohol (ethanol) vapors rise upward, while the remaining wash cascades downward, touching each subsequent heated plate along the way.

While the wash travels down the length of the analyzer, its desirable spirit vapors mix with steam, while some of its impurities rise, and are siphoned to the top of the analyzer. The waste is released at the bottom of the analyzer and discarded.

↓

The hot vapor-steam mixture siphoned to the top travels along a pipe down to the base of the rectifier column. The rectifier is divided into horizontal sections by perforated plates. The column is hot at the bottom but progressively cooler toward the top.

↓

As the hot vapor-steam mixture ascends section by section in the rectifier, it begins to cool, condensing the steam back into water. While alcohol vapors rise, much of the mixture's water content condenses and falls to the bottom, where it is discarded as the tails.

↓

The mixture reaches a point where alcohol begins to condense. At this point there is a spirit plate, a solid rather than perforated plate that allows the condensing, now-liquid alcohol to collect and be siphoned off.

↓

Only the most volatile vapors—the undesirable heads (e.g., methanol, aldeyhdes, esters)—continue to rise and eventually reach the top to be wasted.

Ideally this entire process yields a high proof distillate—roughly 96 percent ABV or 192 proof—that is nearly void of impurities or congeners.

Methods have evolved since Coffey's time, as distillers constantly pursue innovations in technique and equipment. Many manufacturers use still systems made up of several columns to effectively and repeatedly distill and rectify a mixture in order to increase the purity of the final product. It is not unusual for a brand to boast that their vodka has been distilled three, four, five, or even more times.

DILUTION

As I mentioned earlier, water is another raw material in vodka production. Most producers insist on using the best quality of water in their vodka—and for good reason. Many vodkas are distilled to 96 percent ABV or 192 proof, but most are sold at 40 percent ABV or 80 proof. To this end, a considerable percentage of water must be added to achieve appropriate dilution and proof.

Understandably, water quality ranks a skinny second to base ingredient quality in terms of impact on flavor outcome. Nonetheless, water is significant as a diluent or to reduce the distillate's proof—say, from 192 to 80. Using anything less than pristine, flavor-neutral water will adversely and significantly affect the taste of the spirit.

Most brands place great emphasis on their water. Manufacturers source their water from proprietary wells, protected reservoirs, springs, lakes, glaciers, or pristine mountain run-off. Others claim their source locations are free of any pollution. Some use distilled water, while some rely on local tap water that has been filtered and purified. Regardless of source, water added to the spirit must be free of minerals, impurities, and other contaminants. Otherwise all time, money, and effort spent producing a quality distillate are wasted.

FILTRATION

Faith and Wisniewski aptly observe, "Throughout its history ... vodka has been the object of an underlying tension between those looking for purity at any cost and those looking for positive qualities." Filtration helps the Master Distiller achieve the desired levels of purity and flavor, once the distillation and rectification process has removed the lion's share of congeners. Filtration is largely optional, and some vodka producers decide not to employ filtration beyond the function of eliminating obvious particulate matter. Instead, these producers prefer their product to retain some congeners, which imbue the spirit with flavor.

Throughout the ages, charcoal or carbon, with its highly absorbent nature, has proven to be among the most efficient, widely-used filtration materials. Preference has even developed for charcoal made from specific trees—pine, birch, poplar, and oak, just to name a few—because of their degree of absorbency. Many producers now use activated charcoal or carbon because its absorbency increases markedly when heated to more than 1,000 degrees Fahrenheit. Once distilled and diluted, vodka is typically pumped through a filtration device (sometimes several) filled with activated charcoal. This process scrubs the distillate of any remaining undesirable particulates, leaving only a pure and clean spirit.

Aside from charcoal, filtration materials and techniques span a diverse range. According to Pokhlebkin, early on vodka producers used "felt of various types; woolen, linen or cotton cloth or cotton wool; paper of various thicknesses and densities; river, sea and quarried sand; (even) broken pottery." Recent additions include ground coconut shells, silver, gold, platinum, and diamonds. Different means, same goal. Again, each brand's filtration approach is based on its preference and specifications. Is one material or method better than another? Arguably not, as long as the producer is satisfied that the method is delivering the desired outcome.

MATURATION AND MINGLING

Vodka is for the most part an un-aged spirit. Only rarely does it spend any appreciable amount of time maturing or "resting" between the filtration and bottling processes. Some producers feel that the rigors of distillation are tempered by resting, which allows the water diluent to integrate into or mingle with the spirit. Unlike for dark spirits, resting for vodka does not occur in a wood container but in a stainless steel or another vessel that is incapable of imparting flavor.

There are a few points in the vodka production process where mingling may occur: (1) when mixing several runs or batches in pursuit of overall consistency, and (2) during the resting period. As noted, the goal of mingling is to give the added water the opportunity to harmonize with the distillate to improve the final product.

BOTTLING

Bottling—including bottle size, shape, color, and decoration—has zero influence on the quality of the spirit within. That is not to say though that bottle features do not influence sales. All spirits, including imports, sold in the United States are regulated. The regulations dictate the volume of spirit a bottle can contain. The fifth, or fifth of a gallon, is the most common size, containing just over 25 ounces or 750 milliliters. Vodkas are no exception; they must adhere to the same US rules. Many vodka brands are imported from countries with different volume standards, but these foreign producers comply with US standards because the profits they earn here can be well worth the trouble.

PACKAGING

Vodka is a clear, featureless spirit to behold. Reason enough that the industry places much emphasis on packaging. Although packaging is completely inconsequential to the quality of the vodka inside the bottle, it has the power to convince us otherwise. We buy with our eyes, and we tend to equate attractive packaging with high quality. Remember though, packaging and pedigree are one thing, what is inside can only be judged by you according to your personal preference.

Image plays a huge part in what consumers choose to buy. What image does a type of car we drive or the brand of clothing we wear project or say outwardly about who we are or how we want to be seen? Vodka producers are keenly aware of this factor. They spend millions of dollars studying their markets, developing comprehensive marketing plans, and perfecting the art and science of packaging. Certain visual images click with consumers, making the packaging successful. Take, for example, the iconic bottles of Absolut (which is based on an early Swedish medicinal shape) and Skyy (which is a distinctive cobalt blue). Change the shape or color of these two bottles and the vodka inside will remain unaffected, but will the brands sell at the same rate? I would be surprised if they did.

PRICING

As with packaging, high price is not an indicator of high quality. But other factors can legitimately influence the price upward. Raw materials, for one: Some producers use only organically grown or unique strains of a grain or potato, while others rely on imported base ingredients rather than use those accessible locally. Fluctuation in cost of raw materials is also a factor, and so is the price of shipping. Vodka distilled and bottled in Europe and then shipped to and sold in the United States, for example, might have a higher price tag as some of the costs are passed along to consumers. In addition, some vodkas are just more expensive to produce than others, forcing their producers to set a high price.

Again, in my experience purchase price is not directly or reliably correlated with quality—and quality is subjective. The good news is quality vodka is available at nearly every price point.

CHAPTER FOUR

Drink It In

SIP, MIX, STIR, SHAKE. EVERY VODKA FAN HAS AN ESTABLISHED PREFERENCE for enjoying the spirit. Here I cover all the bases—neat, over ice, straight up, or in a cocktail—giving you thoughts and suggestions to consider along the way.

ENJOYING VODKA AT ITS BEST—NEAT

Not everyone is inclined to sip vodka neat as though it were whisky or brandy. It is simply not a prominent feature of Western drinking mores. But if you've read this far, chances are the idea is at least becoming reasonable. Let me encourage you to give it a try.

The only way to decide on your favorite vodka—apples to apples, one style against another—is to taste it in its unadulterated and entirely exposed state—neat, no ice, no garnish. Foreign practice to many, but for others an outright ritual. Such is the Eastern European custom.

Neat is without doubt the best way to enjoy and admire the beauty of a spirit. Vodka is, after all, comparatively "tasteless" and celebrated for its neutrality. In the absence of outside influences—by way of mixers, modifiers, garnish, even ice— and with a little practice, patience, and careful attention to the senses, vodka's nuanced flavor and aroma become increasingly apparent. Once experienced in this form, neat may very well be your new indulgence.

Freezer Cold versus Room Temperature

Though a matter of opinion, storing vodka in the freezer so that it can be poured and consumed while icy cold has become a ritual for me. The freezer's chill creates a viscosity that is silky, syrupy, and creamy on the palate. Serving glasses also should be kept freezer-chilled to retain the spirit's cold temperature. Stored and served similarly, other spirits don't deliver the same kind of pleasing sensation.

When drinking, take your time. Vodka should be sipped and savored rather than shot back in a flash. As its temperature rises, the nose, palate, and mouth feel transform. During the transition from freezer cold to room temperature, vodka's flavors expand, subtle aromas escape, and texture gradually shifts from syrupy to a natural mouth feel. It is this "reveal" that should not be missed, as it affords a dynamic experience of the spirit's character and range.

The recommendation is different, however, in blind tastings. Vodka should be compared neat and at room temperature because it is impractical for testers to wait for the reveal. The full character of its taste, aroma, and texture is present at room temperature, not when freezer cold. Save the pleasure of pouring icy vodka and savoring it at room temperature for when you have selected one or two favorites.

FRESH INGREDIENTS

Let's face it, vodka as a sipping spirit is light-years away from becoming commonplace in the West. Cocktails represent a lion's share of the vodka we consume—one in four, at least. Having plowed headlong into vodka's domain, rest assured I will continue to promote its virtues and the pleasure of enjoying it neat. As a bartender, however, I cannot overestimate the versatility of vodka as a platform for mixed drinks and cocktails. It is an enormous and a vital asset for me, as I love to create fresh-fruit and juice-centric drink recipes.

Vodka's flavor and character nuances, subtle though they may be, can enhance the cocktail experience. Light and citrus forward versus spicy and bold—there is a place for each in creating an outstanding cocktail. That is, as long as quality spirit is mixed with fresh ingredients, end to end. As with most things in life, a product is only as good as its weakest component; quality matters. My career has essentially revolved around two simple principles: (1) fresh ingredients are vital and (2) quality cannot be imitated. Results will always betray practice. Hence use the best possible ingredients when making your drinks, including the spirit and anything that accompanies it.

A considerable challenge in the industry is the influence of profits over ideal practice. The ultimate goal of any spirit producer is to manufacture a superior product, but when doing so threatens the bottom line the intention to deliver the highest quality can become secondary. Using only fresh and top-level ingredients is expensive and cuts into profits, so it can represent a challenging business model for producers. The same paradigm exists in the bar setting. Professional bartenders must balance their intention to serve up the most amazing cocktail experience with the reality of having to use spirits and ingredients that are less than premium. This is a constant struggle, and it takes a great deal of dedication and talent to keep the latter from overcoming the former. Fortunately in the past 10 to 15 years the culture of mixing drinks made a turn for the better. Use of fresh ingredients and superior or boutique spirits has become a movement, one that consumers are flocking toward.

Fortunately, for mixing drinks at home the issue of profit is replaced by time and affordability. I don't support drinking more, but I do promote drinking better. Research and then select top-shelf spirits; remember, quality does not always mean pricey. Pay the extra cost if necessary, though; it is, after all, a treat, and the results won't disappoint. Likewise, spend time, energy, and coin on fresh fruits or savories of the season. Whether juicing; pressing; squeezing; muddling; garnishing; or making a syrup, purée, or infusion, don't even bother to arm wrestle Mother Nature. If the fruit is not at its peak and you wouldn't eat it fresh from the grocer's, don't mix it in your drinks or use it as a garnish. Set your own bar high, and your cocktails will be absolutely amazing.

THE BEAUTY OF ICE

Pristine or premium ice is, like fresh seasonal fruits and quality spirits, essential to assembling the best possible result in a glass. Just as spirit producers should use nothing but clean, pure, flavor-neutral water to dilute their distillate, so should you not add poor-quality ice into your mixed drinks, or the results will be disastrous.

An unwritten ingredient in nearly every mixed drink is water. It is a factor all bartenders consider when preparing cocktails with ice or working on a drink recipe. Ice thrown into the mixing glass or shaker adds as much as 20 to 25 percent water into the drink's content. This dilution is an important consideration in cocktail preparation, and so is the state of the ice. Imagine the impact of ice tainted with freezer burn, high levels of chemicals, and other impurities. You will have delivered a cocktail that extracts 20 percent of its flavor from something you wouldn't drink on its own. Consider a vodka Martini straight up—my recipe, that is (page 94). Here nothing impresses the palate but the vodka and the ice, so the impact of each is even greater. Vodka is remarkably nuanced. Adding ice that delivers unwanted flavor ruins the experience.

Unfortunately, most home freezers do not produce what I call "premium ice"—a problem primarily caused by the water source in a region. Hard or chlorinated water that flows from the tap translates to poor ice. Short of installing a pricy filtration system, the best alternatives are to make ice using distilled water or to purchase ice from the store. When making ice, use silicone ice trays that yield cubes about 1½ inches square. A single large cube (or two) leaves less surface area than a collection of small cubes; this helps prevent over-dilution. Once the cubes are frozen solid, remove from the trays, place in Ziploc-style storage bags, and leave in the freezer until needed. Don't wait too long before use. Covered or uncovered ice idling in the freezer for long periods absorbs the smells and tastes of whatever surrounds it.

My simple message bears repeating: Premium ice is essential to assembling the best possible result in a glass. Quality begets quality—that's the cornerstone and perhaps the single best piece of mixology wisdom. From quality spirit, fresh ingredients, and premium ice, an amazing cocktail is eager to emerge.

GLASSWARE

I argue that glassware is much more than a utensil. Your efforts to prepare, serve, and enjoy a superbly crafted tipple are sorely diminished by a plastic tumbler or a disposable cup. A well-balanced cocktail deserves an elegant glass vessel. Enough said.

VODKA COCKTAILS: NEW FAVORITES AND FAVORITES REVISITED

Are you ready to mix a vodka-based cocktail?

Following are a selection of vodka recipes—some original, some classics, and some with my personal twist. Use them as is or as an inspiration for designing a vodka tipple of your own. Any quality vodka of relatively neutral character works beautifully in all of these recipes. I do, however, offer suggestions of particular vodka types that will enhance the recipe's flavor combination. But do change them up and see which one works best—as always, your preference rules. Remember, with high-quality building blocks, it's pretty tough to fail.

[Unless specifically noted otherwise, each of these recipes makes one drink or cocktail]

APPLE BLOSSOM

With the Apple Blossom, Tony has taken the ordinary and made it extraordinary. He looked at the neon-green, Kool-Aid–like cocktail called the Apple Martini and transformed it into a drink that is destined to be a classic.

DALE DEGROFF,
MASTER MIXOLOGIST
AND AUTHOR

As the craze for all things Martini grew to epic proportions, this cocktail was created to bridge the gap between the popular Sour Apple Martini and the demands of a more sophisticated palate. The Apple Blossom is an updated, thoughtful interpretation with culinary sensibilities. It preserves the spirit of the Sour Apple but takes the concept farther—to one that greets the palate as an honest-to-goodness Granny Smith apple rather than a green-flavored Lifesaver candy. Think of it as a wonderful partnership between tart acidity and natural sweetness. Of course, there will always be a place for the familiar, tongue-zinging Sour Apple tipple. The Apple Blossom is a refined and balanced upgrade.

Using a vodka with lots of acidity— something featuring fruit and citrus, such as those made of wheat or mixed grains—works great.

1 oz (30 ml) vodka
½ oz (15 ml) calvados
1 oz (30 ml) freshly squeezed lemon juice
½ oz (15 ml) simple syrup (see page 112)
1 oz (30 ml) green apple purée (see page 112)
1 tbsp egg white

In a mixing glass add vodka, calvados, fresh lemon juice, simple syrup, green apple purée, and egg white. Shake with ice until well blended; strain into a chilled cocktail coup. Garnish with three drops of Peychaud's bitters.

BLACK BOOT

They're professionals at this in Russia, so no matter how many Jell-O shots or Jäger shooters you might have downed at college mixers, no matter how good a drinker you might think you are, don't forget that the Russians—any Russian—can drink you under the table.

ANTHONY BOURDAIN

One of the many drinks that gained popularity during the Disco era, the White Russian has gained a new cult following—thanks to its appearance as the Dude's favorite drink in the 1998 film *The Big Lebowski*. According to Dave Wondrich, "When [he] first encountered it in the 1970s, the White Russian was something real alcoholics drank, or beginners." Now, ordering the drink is "the mark of the hipster." Credit has been given to Donato (Duke) Antone—creator of the Harvey Wallbanger—and Duke's Blackwatch Bar for coming up with the White Russian, though this claim remains a little tenuous for me.

The Black Russian, meanwhile, is a simple mix of vodka and coffee liqueur, created in the Hotel Metropole in Brussels. Barman Gustave Tops is said to have fashioned it in 1949 for Perle Mesta, the US ambassador to Luxembourg. Easy on the taste buds and a favorite of those with a sweet tooth, the Black Russian is among the easiest mixed drinks to pull together. May I suggest this bumped-up version I call the Black Boot?

The Black Boot sits best with an Old-World, Russian-style vodka—something with assertive character and pronounced spice, cocoa, and vanilla notes.

2 oz (60 ml) vodka
1 oz (30 ml) Kahlúa
½ oz (15 ml) Ramazzotti Amaro

In an ice-filled mixing glass add vodka, Kahlúa, and Ramazzotti. Stir until well blended. Strain into an ice-filled Old Fashioned glass. Serve with a swizzle stick.

WHITE RUSSIAN

2 oz (60 ml) vodka
1 oz (30 ml) Kahlúa
1 oz (30 ml) heavy cream

In a mixing glass add vodka, Kahlúa and heavy cream; shake with ice until well blended. Strain into an ice-filled Old Fashioned glass. Serve with a swizzle stick.

BLOODY BULL

I remember the drink, but not who invented it.
I remember them joking about drinking soup.

ELLA BRENNAN

As a vetted member of the American cocktail lexicon, the Bloody Mary has inspired a number of delicious and enduring relatives, my favorite being the Bloody Bull. With origins that are elusive at best, several trails lead to Brennan's, one of New Orlean's most influential culinary houses known for delivering a dining experience where patrons should expect to either go big or go home. The addition of bouillon to a familiar ensemble of tomato, lemon, Worcestershire sauce, and the rest of the recipe's "Bloody companions" gives this drink a hearty and distinctly food-friendly character with a good measure of gusto. Perfect alongside traditional, full-on savory breakfast fair—Corned Beef Hash, Eggs Benedict with its hallmark hollandaise sauce, Oysters Rockefeller, the list goes on. All of these would be dutifully served by a well-made Bloody Mary, no question. But partnered with a Bloody Bull, the entire breakfast experience is kicked to a new level—rich, complex, and complementary without commandeering the palate.

In this one a spicy, robust vodka—something made from rye or mixed grain—is best. Try substituting the beef bouillon with elk if you are up for a little adventure. It yields a strong, savory character—a real hit with the crowd when it was unleashed at the Aspen Food & Wine Classic.

2 oz (60 ml) vodka
2 oz (60 ml) tomato juice
2 oz (60 ml) beef bouillon
½ oz (15 ml) freshly squeezed lemon juice
3 dashes Worcestershire sauce
2 dashes Tabasco sauce
Pinch Kosher salt
Pinch coarsely ground black pepper

Place all ingredients into a mixing glass, add ice, and roll contents between mixing glass and shaker tin until well mixed. Strain into an ice-filled Collins glass. Garnish with a wedge of lemon.

CAESAR

He's Italian and a fantastic cook and came up with a concoction of mixing clam juice, with tomato juice and Worcestershire sauce, Tabasco and vodka and because of his Italian ancestry, decided to call it Caesar.

SHEENA PARKER, GRANDDAUGHTER OF CAESAR'S CREATOR

Canadians will recognize this savory tipple as their own. In 1969, Walter Chell created the Caesar, sometimes dubbed the Bloody Caesar, at the Calgary Inn, where it was featured in the hotel's new Italian restaurant. Perhaps the only cocktail recipe inspired by a pasta dish—alle vongole (spaghetti with tomato sauce and clams) that Chell enjoyed on a trip to Venice—the Caesar has unique underpinnings: a spicy mix including mashed clams and tomato juice. Around the same time, the Duffy Mott Company introduced Clamato, a combination of clam and tomato juice, rescuing countless bartenders from having to prep and handle mashed clams behind the bar. Fine in a pinch, but scratch is always best—use tomato juice and clam juice separately to mimic Chell's original. This drink was hugely popular at the Brass Rail, where I first tended bar in 1980. Located in Port Huron, Michigan, just across the Blue Water Bridge from Canada, we served a sturdy clientele of Caesar fans.

This is a feisty drink. Respect its big flavors and reach for an Old-World, potato-based or rye-based vodka—something with a little spice to boost the flavor. Have fun with the garnish; almost anything can adorn the modern Caesar. Pickles, olives, peppers, carrots, or spicy pickled string beans—virtually any pickled vegetable is a nice addition.

2 oz (60 ml) vodka
½ oz (15 ml) freshly squeezed lime juice
2 dashes Tabasco sauce
3-4 dashes Worcestershire sauce
Pinch Kosher salt
Pinch coarsely ground black pepper
3 oz (90 ml) tomato juice
2 oz (60 ml) clam juice

Place all ingredients into a mixing glass, add ice, and roll contents between mixing glass and shaker tin until well blended. Rim the lip of a double Old Fashioned glass with lime and dip into a mixture of cracked black pepper and celery salt. Strain into the prepared, ice-filled double Old Fashioned glass. Garnish as desired.

CAIPIROSKA

Look at me and tell me if I don't have Brazil in every curve of my body.

Carmen Miranda

Visiting Rio de Janeiro and my friend Carlos Alves is always an honor. His family owns and operates Quiosque do Português, one of the busiest quiosques (bars) along the iconic beaches of Copacabana, Leblon, and Ipanema. Last visit, as usual sitting and quenching my thirst with a Caipirinha—Brazil's cachaça-based national drink—I was struck to see most customers ordering Caipiroska, which in essence is the same muddled cocktail, but with vodka in place of cachaça. According to Carlos, vodka is gaining steam as a rival to Brazil's long-standing national spirit.

The Caipiroska follows a simple recipe of fresh lime, sugar, and vodka. Debate exists as to what type of sugar—granulated white or simple syrup. An entirely personal preference to my eye, though from my travels granulated seemed universally used. Regardless, this drink is a fantastic platform for experimentation. Consider the array of alluring seasonal fruit combinations out there. I tend to keep the classic recipe mostly intact by keeping fresh limes in the mix, then add whatever is seasonally fresh and readily available—strawberry, blueberry, clementine, kumquat. Select one or try a combination; it's tough to go wrong.

A great match is a clean, neutral vodka with fruity, floral notes—something made from corn, wheat, or mixed grain, something gentle and that complements the drink's citrus element.

2 oz (60 ml) vodka

1 oz (30 ml) simple syrup (see page 112) or 1 heaping tbsp (approx 4 tsp) granulated white sugar

1 small lime cut in quarters

Fill an Old Fashioned glass with cracked ice. In a mixing glass add simple syrup or granulated sugar and lime quarters. Muddle to extract juice without forcing the rind from the lime. Dump the ice from the glass into the mixing glass, add vodka, and shake. Pour the entire drink into the chilled Old Fashioned glass. Remember, this drink is served with the same ice used during its preparation.

COCOA À TROIS

All you need is love. But a little chocolate now and then doesn't hurt.

CHARLES M. SCHULZ

Thrown together during the filming of *Giant* in the dusty Texas town of Marfa, film legends Rock Hudson and Elizabeth Taylor became fast friends. Challenged as they were for after-hours entertainment, and young enough to endure the indulgence gracefully, the two decided to concoct a drink of chocolate, vodka, and Kahlúa. They called it the Chocolate Martini. According to a Rock Hudson fan blog, Elizabeth Taylor boasted, "during our toots, we concocted the best drink I ever tasted—a chocolate Martini, made with vodka, Hershey's syrup and Kahlúa. How we survived I'll never know." Rock Hudson reminisced, "We were really just kids, we could eat and drink anything and we never needed sleep...." Where was James Dean, I wonder?

Though technically not a Martini, the actors' drink inspired countless other cocoa-centric tipples, a desert-type liquid indulgence for chocolate lovers around the globe. In honor of Rock Hudson and Elizabeth Taylor's liquid adventures I offer Cocoa à Trois—a yummy libation featuring chocolate three ways.

Undeniably dessert in a cocktail glass. Given its rich chocolaty underpinnings, the drink will pair best with a New-World corn-based or wheat-based vodka rich in vanilla and cocoa elements.

2 oz (60 ml) vodka

1 oz (30 ml) Godiva Chocolate liqueur

½ oz (15 ml) simple syrup (see page 112)

¼ tsp (2 ml) sweetened cocoa powder

Grated semisweet chocolate for garnish

1 tbsp (15 ml) egg white

In a mixing glass add vodka, Godiva liqueur, simple syrup, sweetened cocoa powder, and egg white. Shake vigorously until blended (shake long enough to dissolve the cocoa powder). Strain into a small (Nick and Nora–sized), chilled cocktail glass and dust with semi-sweet chocolate.

COSMOPOLITAN

Far from a malevolent symbol of hyperentitled wretched excess, it is actually a benignly simple and, I still must say, fetchingly balanced drink.

TOBY CECCHINI, BARMAN

Visible as a bar-menu mainstay starting in the late 1980s, the Cosmopolitan remains firmly at the head of the pack, best known among a small group of new classics. It belongs to what Gary Regan in the *Joy of Mixology* refers to as the New Orleans Sours—which consist of a base spirit, orange liqueur, and fresh citrus—a collection that includes "some of the world's greatest cocktails … the Side Car, the Margarita, and the Cosmopolitan among them." Barman Toby Cecchini feels that this group's unique appeal is due to its "triangulation of liquor, which gives body and punch, with no added frippery, that even this many years later [he] has not been able to improve."

Who created the "Cosmo" is still up for grabs. But leave it to an icon to catapult the drink to star status—once Madonna was spotted sipping one, its fate was sealed. The Cosmopolitan was ordered coast to coast almost overnight. For the younger set, fans of *Sex and the City* served as a conduit, bringing the Cosmopolitan into our viewing consciousness and further securing its rank as one of the most requested cocktails in the world.

In keeping with celebrating vodka at its purest level, flavored vodkas were intentionally excluded from this book. That said, the Cosmopolitan is a cocktail for which I might reach for a citrus-flavored vodka. Here, I suggest vodka with a lot of bright citrus notes in both nose and palate—something approachable, made of grape or barley. Or better yet, taste it both ways, using flavored and unflavored vodka side by side, and see what you think.

1 ½ oz (45 ml) vodka
¾ oz (22.5 ml) Cointreau
½ oz (15 ml) freshly squeezed lime juice
¾ oz (22.5 ml) Ocean Spray cranberry juice cocktail

In a mixing glass add vodka, Cointreau, lime juice, and cranberry juice; shake with ice until well blended. Strain into a chilled cocktail glass and garnish with a spiral of lemon.

CUCUMBER COBBLER

Without doubt the Sherry Cobbler is the most popular beverage in the country, with ladies as well as with gentlemen.

HARRY JOHNSON

Cobblers found their stature in the latter part of the 19th century. Roughly defined, they are made of wine or spirit, sugar, and fresh fruit; they are shaken, served in a tall ice-filled glass, decorated with fresh fruit, and served with straws. The most popular Cobbler back in the day was the Sherry Cobbler—fresh orange slices, a little sugar, and a lot of sherry—decidedly sweet by today's standards.

The Cobbler formula remains a great template for experimentation. It may not seem an intuitive choice, but cucumber can be very interesting and refreshing—one that serves to moderate sweet elements. I first saw cucumber as an ingredient in the refreshing and respectable Pimm's Cup. Though a garnish, it delivered considerable flavor and an element of balance. Less novel today as a drinks ingredient, cucumber has a way of bringing a clean, distinctive brightness, elevating the refreshment factor. Think of the clean, bright, thirst-quenching effect of cucumber water served at most spas. It's also great, even unexpected, when combined with other flavors such as citrus, ginger, and apple.

This drink carries real flavor diversity—slightly sweet and fruity with a spicy, savory side. A vodka with a light, somewhat neutral character and notes of spice, citrus, and green apple works best. Generally, I use vodka made of mixed grain.

2 oz (60 ml) vodka

1/2 lemon, sliced

1 oz (30 ml) ginger syrup (see page 112)

2 oz (60 ml) fresh apple-cucumber purée (see page 112)

In a mixing glass muddle lemon slices with ginger syrup; add vodka and apple-cucumber purée. Shake with ice, and strain over cracked ice in a double Old Fashioned glass. Garnish with apple, cucumber, and lemon slices. Straws optional.

ESPRESSO MARTINI

Espresso consumption is an aesthetic experience, like tasting a vintage wine or admiring a painting. It is a search for beauty and goodness and improving the quality of our life.

ANDREA ILLY

The mid to late 1980s witnessed a Martini explosion. The word "Martini" was attached to any drink served in a cocktail glass—popularized into what is commonly considered a Martini glass. Martini bars materialized everywhere. New drinks, served in the aforementioned cocktail glass, were being created in droves and were available in every flavor and color imaginable. Even established drinks—the Cosmopolitan and the Lemon Drop, among them—were mistakenly given the Martini moniker. Notably, gin was nowhere in sight. The vodka floodgates had fully opened.

The decades since have passed with numerous Martini-esque fledglings lost along the way, yet the Espresso Martini has flown the distance. Bartender and London legend Dick Bradsell is known to be its alchemist; his was known as the Vodka Espresso. In 1984 he was behind the stick at Fred's Bar in the Soho neighborhood in London when a beautiful model asked for a drink "that would wake [her] up then @$%! [her] up." And so it was done, and the Espresso Martini endures.

Lots of great Espresso Martini renderings are out there, but the following recipe is what I enjoy the most. Because I love the combination of vanilla with coffee, I reach for a vodka with notes of vanilla, cocoa, and toffee—something of a winter wheat and New-World character.

1 ½ oz (45 ml) vodka
¾ oz (22.5 ml) Kahlúa
1 oz (30 ml) chilled, freshly brewed espresso
1 oz (30 ml) heavy cream
½ oz (15 ml) simple syrup (see page 112)

In a mixing glass add vodka, Kahlúa, chilled espresso, heavy cream, and simple syrup; shake vigorously until well blended and frothy. Strain into a chilled cocktail glass. Garnish with a sprinkle of ground espresso and ground cocoa.

FLAME OF LOVE

I'd hate to be a teetotaler. Imagine getting up in the morning and knowing that's as good as you're going to feel all day.

DEAN MARTIN

One of the great privileges of my career has been working with so many industry greats—in particular Dale DeGroff, a mentor and inspiration to many. Dale and I traveled extensively on behalf of Finlandia— a unique opportunity to focus heavily on the finer points of vodka. The Flame of Love was among Dale's featured cocktails at these events. Dale's storytelling was all the more entertaining, for he knew Pepe Ruiz, the drink's architect.

Pepe, a barman at the 1970's celebrity hangout Chasen's in West Hollywood, was challenged by the infamous Dean Martin to come up with a special Martini. Enter La Ina Fino sherry and a technique for imparting orange essence I have come to love— flamed orange peels. Upon Dean Martin's next visit to Chasen's, Pepe greeted him with, "Mr. Martin, your Flame of Love Martini!" Enjoyed, no doubt, to the last drop.

La Ina Fino is a delicate sherry with flavors of ripe apples, almonds, and vanilla. It works beautifully alongside the floral essence of orange delivered by a flamed peel finish. For this one, I might use a vodka made of either barley or maple sap—the former is strong with green apple notes, while the latter shows prominent vanilla and almond tones.

3 oz (90 ml) vodka

½ oz (15 ml) La Ina Fino sherry

3 peels of orange for flaming

Rinse a chilled cocktail glass with La Ina Fino sherry and discard the excess. Flame two orange peels over the glass to coat the inside with the oils. Chill the vodka well over ice—Pepe shakes it; Dale stirs it. Strain the drink into the prepared glass, and garnish with a flamed orange peel.

GIMLET

The bartender set the drink in front of me. With the lime juice it has a sort of pale greenish yellowish misty look. I tasted it. It was both sweet and sharp at the same time. The woman in black watched me. Then she lifted her own glass towards me. We both drank. Then I knew hers was the same drink.

RAYMOND CHANDLER,
THE LONG GOODBYE

Great Britain's 1867 Merchant Shipping Act mandated daily rations of lime or citrus juice for every sailor in its Naval and Merchant fleets. (Citrus was known to combat the devastating problem of scurvy, a potentially lethal result of vitamin C deficiency that threatened the fleet's stability.) That same year Lauchlin Rose—a member of a shipbuilding family that fully appreciated the seafarers' scourge—patented the process for preserving citrus juice without alcohol, creating Rose's Lime Juice. Years later British Royal Navy Surgeon General, Sir Thomas D. Gimlette was credited for introducing the mixture of Rose's and gin as an effective means of persuading sailors to consume their "limey" rations—a combination that came to be called the Gimlet.

Many gin-based cocktails have evolved over time— Martini a prime example—but the Gimlet has become a vodka-based drink, defined primarily by its sweetened lime cordial. As Gary Regan in *The Joy of Mixology* maintains, Rose's Lime Juice "is an integral part of ... the Gimlet." *Preserved* lime juice, you ask? It does rub against the fresh ingredients principle I advocate, but I use Rose's in this drink without reserve. I also recommend serving a Gimlet with a wedge of *fresh* lime to allow the option of adjusting the drink's sweetness with just a squeeze.

In order to balance Rose's acidity I use a vodka with lots of character and a rich mouth feel—potato based, perhaps— something with a rich, creamy texture.

2 oz (60 ml) vodka

1 oz (30 ml) Rose's Lime Juice cordial

In a mixing glass add vodka and lime juice; shake with ice until well blended. Strain into an ice-filled Old Fashioned glass. Garnish with a wedge of lime.

GLÖGG

Celebrate the happiness that friends are always giving, make every day a holiday and celebrate just living!

Amanda Bradley

love the holiday tradition of sharing a festive tipple with friends and family. A big bowl of punch, a steaming cup of Hot Buttered Rum, a glass of egg nog, or a mug of Tom and Jerry. It's no wonder every household, indeed culture, has a holiday tradition; the holidays are simply better celebrated with a cherished seasonal beverage.

My dear German friends Walter and Christa introduced me to Glühwein, their native version of mulled wine—made with red wine, spices, sugar, and water. It is quite delicious and always warms my body and soul. The main difference between Glühwein and its Nordic cousin Glögg is that the latter has vodka and the rich complexity delivered by the addition of raisins to the mix and its nutty garnish. Different from your typical mulled wine, but Glögg is an amazingly delicious brew.

When choosing a vodka for Glögg I pay attention to tradition and select one from the Nordic producers. Sweden, Finland, Norway, Denmark, and Iceland all produce wonderful options.

8 oz (240 ml) vodka
1 bottle (750 ml) dry red wine
½ cup (120 ml) golden raisins
⅓ cup (107 ml) sugar
Peel of a small orange, peeled as you would an apple
1 stick cinnamon
6 whole cloves
2 cardamom pods
¼ cup (60 ml) blanched, sliced almonds for garnish
¼ cup (60 ml) golden raisins for garnish

Stir together vodka, wine, raisins, and sugar in a large pan. Place orange peel, cinnamon, cloves, and cardamom in a square of cheesecloth and tie into a bag. Add spice bag to wine mixture. Heat to a simmer, uncovered for 10 minutes. Do not boil. Remove and discard the spice bag. Serve in heated mugs; garnish with sliced almonds and golden raisins.

GYPSY QUEEN

All the smart bars here [in New York] are now [in 1934] serving vodka, and many of the accomplished drinkers are quaffing it in lieu of their favorite tipple.

O.O. McIntyre

Unlike gin, which is present in a long list of drink recipes that stretch back to at least the early 1800s, vodka is a relative newcomer as a mixed drink feature. In truth, many of the earliest mixed drinks are of a style, blend, and balance that pair less favorably with the contemporary palate. Often just too sweet or overpowering. One of the vintage cocktails I enjoy, however, is the Gypsy Queen. Created at the Russian Tea Room, it is found in the establishment's 1938 publication, *Russian Dishes and What They Are Made Of.* I agree with Dave Wondrich, who says, "This was their most famous drink—and deservedly so, for it demonstrates one of vodka's greatest mixological strengths, its ability to smooth out strong flavors." Bénédictine, this drink's ingredient, is as formidable, strong, and complex as they come.

For this, I like a vodka with bold, assertive character—Old-World, mixed-grain Russian with baking spice features. If you have not yet had the opportunity, treat yourself to a little sip of Bénédictine—neat. It is a flavor journey that is better experienced than described, as mere words do it a disservice. *Deo Optimo Maximo* (D.O.M.)—for our best, greatest God—just a sip of it and you'll see why.

2 oz (60 ml) vodka
1 oz (30 ml) D.O.M. Bénédictine
2 dashes Angostura bitters

In an ice-filled mixing glass add vodka, Bénédictine, and Angostura bitters; stir until very cold. Strain into a chilled Old Fashioned glass. Garnish with a thin slice of lemon peel.

H2O COCKTAILS

Water is the driving force of all nature.

LEONARDO DA VINCI

Vodka has traditionally been celebrated in the West for its neutrality, and remains largely relegated to serve the role of a drink's foundation—merely supporting the flavors added to it. To enjoy vodka for its own set of flavor nuances, we offer a new approach to showcasing the spirit: the H2O Cocktail.

H2O Cocktails deliver a chic, sophisticated, fresh, and bright solution. Delicately flavored infused waters are mixed with vodka, enhancing and complementing the spirit's nuances rather than burying them deep within.

Thank you to my fellow mixologist Kathy Casey for introducing me to these unique and innovative tipples—they are the perfect vehicle for showcasing vodka's assets.

Once you get your mixing rhythm, your flavored water potential is endless. Fresh fruits, herbs, spices, vegetables, flowers—just go for it! As for vodka selection, choose based on the features you are working to enhance.

MANDARIN HIBISCUS H2O

A slightly floral, fruity cooler with just a touch of spice. It works great with the balanced complexities of a mixed-grain vodka.

1 ½ oz (45 ml) vodka

3 oz (90 ml) mandarin hibiscus water (see page 113)

In an ice-filled Highball glass add vodka and top with infused water. Stir to mix.

CUCUMBER RED BELL PEPPER H2O

A savory blend, slightly herbal with lovely vegetal qualities. It pairs perfectly with potato vodka.

1 ½ oz (45 ml) vodka

3 oz (90 ml) cucumber red bell pepper water (see page 113)

In an ice-filled Highball glass add vodka and top with infused water. Stir, and garnish with a bouquet of fresh cilantro.

BERRY H2O*

Lightened with refreshing berry-infused water. It's slim and sophisticated with a dry berry finish; best with a wheat or mixed-grain vodka.

1 oz (30 ml) vodka

3 oz (90 ml) berry water (see page 113)

In an ice-filled Highball glass add vodka and top with infused water. Stir, and garnish with fresh berries on a pick.

PINEAPPLE CILANTRO H2O*

Fruit and herb notes grace this less-sweet drink. I recommend a mixed-grain, complex-character vodka.

1 oz (30 ml) vodka

3 oz (90 ml) pineapple cilantro water (see page 113)

In an ice-filled Highball glass add vodka and top with infused water. Stir, and garnish with a small, thin slice of pineapple or cilantro sprig.

*COURTESY OF KATHY CASEY

HARVEY WALLBANGER

Disco music in the '70s was just a call to go wild and party and dance with no thought or conscience or regard for tomorrow.

Martha Reeves

The first time I encountered "Harvey" was when I was ten or eleven in the early 1970s. My older cousin, Stan Carroll, who was one cool cat, had a tee shirt adorned with a surfer character named Harvey Wallbanger. At the time I wasn't aware of the drink reference to the now legendary bold combination of vodka and orange juice with a Galliano float. According to William Grimes in *Straight Up or On the Rocks*, the birth of this drink coincided with Galliano producer's major advertising campaign, which featured the Harvey character.

The story was that Harvey was bereft at having lost a big surfing competition, so he imbibed one too many drinks made by his local Hollywood bartender Duke Antone. In a state of inebriated frustration, Harvey banged his head against a wall—hence the decidedly catchy name. Truth or tale? Regardless, the story caught on and led to the drink's popularity, tripling the sales of Galliano.

Created in 1896 by Arturo Vaccari in Livorno, Italy, Galliano is a liqueur flavored with vanilla, cinnamon, star anise, lavender, and ginger, among other botanicals. It is a great match for a New-World mixed-grain vodka with both floral and spice notes including white pepper, cinnamon, clove, and vanilla.

2 oz (60 ml) vodka

4 oz (120 ml) freshly squeezed orange juice

½ oz (15 ml) Galliano

In an ice-filled Collins glass add vodka and orange juice; stir well to mix. Float Galliano on top and garnish with a spiral of orange and a swizzle stick.

HURLYBURLY

When the hurlyburly's done, when the battles lost and won.

SHAKESPEARE, *MACBETH*

Hurlyburly is the name of a stage production tackled by an enthusiastic but somewhat ragtag company of actors—the San Francisco–based Pour Boys—in 1997. Hurly-burly refers to a state of chaos, confusion, and turmoil. That pretty much sums up those days for me and many of my fellow thespians—all starving actors/bartenders who frequently hosted parties to raise funds to finance the cost of our stage productions. I developed this drink just for such an occasion. With a donated case of vodka on hand, I made something fun, fruity, whimsical, and easy to drink. Everyone loved it.

Unlike our play, this cocktail is not meant to be taken too seriously, just enjoyed. It follows the same formula as the Cosmopolitan, Side Car, White Lady, Margarita, and Cable Car—each made with the divine trilogy of base spirit, orange liqueur, and citrus.

This drink is perfect for a fun, lightly citrus vodka with notes of vanilla and high acidity. Something New World made from winter wheat or corn should fit the bill nicely.

1 ½ oz (45 ml) vodka

½ oz (15 ml) Cointreau

½ oz (15 ml) Ocean Spray cranberry juice cocktail

½ oz (15 ml) freshly squeezed lemon juice

½ oz (15 ml) freshly squeezed orange juice

In a mixing glass add vodka, Cointreau, cranberry juice, fresh lemon and orange juices; shake with ice until well blended. Strain into a chilled cocktail glass and garnish with dried cranberries.

INFUSIONS

We are in the era of Mixology. They do wonderful things with vodka infusions, blending with the amazing flavours that we can get today. Vodka goes so well with so many ingredients .... Vodka is obviously a prime spirit.

PETER DORELLI

Infusing or flavoring vodka goes back nearly as far as vodka itself. Fruits and herbs weren't originally used to enhance flavor; instead they served to mask unpleasant congeners left behind from unrefined distillation methods. In time the necessity to mask a questionable distillate developed into a highly specialized segment of vodka's legacy. There is an abundance of quality flavored vodkas serving invaluably behind the bar. Creating infusions at home, however, is a particularly fun and relatively easy way to add flavor options not available on market shelves.

Use infusions to develop new recipes or to augment those you already enjoy. Fruits, herbs, vegetables, and spices—select one or combine several into a creative concoction. Let your imagination and culinary senses prevail. Love citrus? Pick from sliced lemons, limes, oranges, clementines, mandarins, and blood oranges and combine two or more to create an original citrus-infused vodka. Prefer spicy? Infuse chilies and peppers of your choice to make a zesty foundation for any Caesar, Bloody Mary, Bloody Bull, or other savory tipple. Try a little crossover—cilantro and mango, chilies and lime, cucumber and ginger, pineapple and vanilla. You get the idea.

One of my recent favorites is a combination of pineapple and vanilla bean infused in a New-World corn-based vodka—approachable, neutral, and a great canvas for flavors. And it makes an amazing twist on a Vodka Sour—fresh lemon juice, simple syrup and egg white.

PINEAPPLE SOUR

2 oz (60 ml) pineapple vanilla bean infused vodka (see page 112)
1 oz (30 ml) freshly squeezed lemon juice
1 oz (30 ml) simple syrup (see page 112)
1 tbsp egg white

In a mixing glass add infused vodka, fresh lemon juice, simple syrup, and egg white; shake vigorously with ice until well blended and egg white is completely emulsified. Strain into an ice-filled double Old Fashioned glass and garnish with a pineapple spear.

KAMIKAZE

*ka·mi·ka·ze [kah-mi-**kah**-zee]—noun.*
A person or thing that behaves in a wildly
reckless or destructive manner.

DEFINITION FROM **DICTIONARY.COM**

A product of the carefree, "unsophisticated" days of the 1970s, this drink is among those largely considered to be a means to an end. Look to a gent named Liam, bartender at Boston's Eliot Lounge. He created a drink of vodka, triple sec, and Rose's lime juice that he called a Liam. His recipe gained rapid success and was swiftly adopted by others in the industry. A few years after Liam's handiwork, the same drink appeared in a Smirnoff advertisement as the "Kamikaze." By the early 1980s, it was all the rage.

Aptly named for the effect it imparted, the Kamikaze was initially a sweet, candied shooter often consumed in multiples and in quick succession—never intended for sipping or savoring, or for the faint of heart. Liam's recipe is exactly how I remember learning to make this drink. I confess to having enjoyed a few myself during my fledgling, carefree drinking days. Fast-forward—I strongly recommend savoring this grown-up version with Cointreau in place of triple sec and fresh, hand-extracted lime juice in place of Rose's.

Anyone wishing to relive their misspent youth can split this recipe into two frozen shot glasses and share with a friend—old-school shooter style. My Kamikazes are made with a very approachable vodka—something New World and corn based with a grainy sweetness.

2 oz (60 ml) vodka

1 oz (30 ml) Cointreau

½ oz (15 ml) freshly squeezed lime juice

In a mixing glass add vodka, Cointreau, and fresh lime juice; shake with ice until well blended. Strain into a chilled cocktail glass and garnish with a spiral of lime.

LEMON SPOT

There was a time, kiddies, when you could go home with a paralegal or a stockbroker for the night and no one had to exchange medical histories (though they may have had to come clean about being a Capricorn).

SAN FRANCISCO CHRONICLE
REPORTED IN A 2004 PIECE
ON PERRY'S FERN BAR

The Lemon Drop's modest beginnings are credited to the 1970s and Henry Africa, an enormously popular fern bar and the brainchild of Norman Hobday. An innovator known for establishing the San Francisco fern bar trend, Hobday was largely responsible for the bar's success in the city's enthusiastic singles scene. The Lemon Drop—aptly named for its resemblance to the familiar lemon-flavored, sweet-and-sour candies rather than its mouth-puckering effect—became a fast favorite. I learned to make Lemon Drops at the Brass Rail in 1981, coached by Tony, my cousin and bartending idol at the time. Back in the day, it meant turning out a chilled shot of vodka served with a granulated sugar-dredged slice of lemon. The approach was to suck the sugary lemon before downing the vodka—inelegant and effective, and to my recollection always fun!

While the '70s' original survived the cocktail revolution that followed, I prefer this more refined interpretation called the Lemon Spot. The addition of Tuaca and its vanilla tones along with orange-flavored liqueur elevates the flavor dimension.

There are lots of flavors to play off of in this recipe. I like using a vodka that veers toward toffee, butterscotch, or caramel notes to complement Tuaca's strengths—something made of corn or mixed grain.

1 ½ oz (45 ml) vodka
½ oz (15 ml) Cointreau
½ oz (15 ml) Tuaca
1 oz (30 ml) freshly squeezed lemon juice
3 or 4 fresh rosemary leaves
1 tbsp (15 ml) egg white

Rim a chilled cocktail glass with superfine sugar. (To sugar frost a cocktail glass, simply moisten the lip of the glass with a wedge of lemon and dip the glass into a plate of lemon zest mixed with superfine sugar.) In a mixing glass add vodka, Cointreau, Tuaca, fresh lemon juice, rosemary leaves, and egg white; shake with ice until well blended. Strain into the sugar-frosted cocktail glass. Garnish with a spot of lemon zest.

MARTINI

Dr. No: *Medium dry Martini, lemon peel, shaken not stirred.*
James Bond: *Vodka?*
Dr. No: *Of course.*

IAN FLEMING, *DR. NO*

I certainly do enjoy drinking a vodka Martini, although I wish we could have named it something different entirely. For tradition's sake, I am more comfortable thinking of a Martini as being made of gin and dry vermouth. What's more—it may be my palate or simply a conditioned habit—I have never really felt that vermouth complements vodka, as vermouth most definitely is a soul-mate to gin. Hence when it comes to vodka Martinis, I generally recommend leaving the vermouth out all together. Also, I prefer mine stirred not shaken.

All this is in direct contrast to the taste of the most iconic Martini drinker of all—007, James Bond. Still, I promote my personal preference. The famous British Secret Service agent's widely known inclination toward a medium-dry vodka Martini that is "shaken, not stirred" influenced an entire generation and beyond to want the same. I encourage you to experiment with both techniques to find yours.

Staying true to my convictions, this recipe is conspicuously sans vermouth, as in my opinion it gets in the way of enjoying vodka's subtle nuances. This point magnifies the importance of paying attention to selecting vodka and garnish. Switch it up to complement the character of your vodka choice. Choose big, blue-cheese-stuffed Spanish olives with spicy rye-based vodka, or pair pearl onions with creamy, rich potato-based vodka; a twist of lemon or orange is perfect with a floral, fruit-forward selection. Consider also that sometimes "naked"—just you and the ice-chilled vodka—can be an exceptionally nice experience; aromatics, flavors, mouth feel, and character slowly reveal, each expression expanding and developing gradually as the temperature rises. If you already enjoy a favorite vodka neat at room temperature, chances are you will love it properly chilled and served in a beautiful cocktail glass.

This Martini above all others supports the benefits of doing a blind tasting. I refrain from making a specific vodka recommendation in hopes you'll take a little time to research, taste a few, identify your style preference, and choose a favorite. Have fun!

3 oz (90 ml) vodka

In an ice-filled (large, hard, dry, cold ice cubes) mixing glass add your favorite vodka; stir or shake until icy cold. Strain into a chilled cocktail glass. Garnish with your favorite accessory.

MONKEY SHINE

May you live all the days of your life.

JONATHAN SWIFT

Tales of the Cocktail is an annual July gathering held in New Orleans, attended by drinks enthusiasts from around the globe. Bartenders, distillers, suppliers, writers, bloggers, fans, and all things cocktail converge in the Big Easy to do what we do and love best—drink and talk cocktails. In 2011, as part of an event called the Cocktail Hour—a gathering of 40 mixologists—the attendees were challenged to craft an original drink inspired by locations around the globe. The Monkey Shine was my entry. Central American–inspired, this aperitif features guava, citrus, and Campari. I love this event—a group of friends and other like-minded spirits geeks exploring new recipes and celebrating the end of a full day in NOLA. What's not to like?!

Guava has a tangy sweetness, is bold but not overpowering, and combines beautifully with a variety of other fruits. Usually I support all tropical flavors with a fruit-forward, clean, approachable vodka—something made of a mix of grains, including barley.

1 ½ oz (45 ml) vodka
¾ oz (22.5 ml) Campari
¾ oz (22.5 ml) Cointreau
1 oz (30 ml) freshly squeezed lemon juice
2 ½ oz (75 ml) pink guava nectar
½ oz (15 ml) simple syrup (see page 112)

In a mixing glass add vodka, Campari, Cointreau, fresh lemon juice, guava nectar, and simple syrup; shake with ice until well blended. Strain into an ice-filled Collins glass. Garnish with mint sprig and lemon fan.

MOSCOW MULE

The nicest thing about the mule is that it doesn't make you noisy and argumentative, or quiet and sullen, but congenial and in love with the world. One wag of its tail and life grows rosy.

CLEMENTINE PADDLEFORD

The Moscow Mule's genesis was the serendipitous relationship between two gents, John G. Martin and Jack Morgan. In 1939 Martin acquired Smirnoff for a song from Rudolph Kunett—possibly because, in part, gin was the white spirit of choice, bar none. Morgan, owner of the Cock 'n Bull restaurant on the Sunset Strip in Hollywood, ventured to import ginger beer to the United States. Faced with the challenge of having to move an overstock of spicy ginger beer and the yet-to-be mainstream Smirnoff vodka, Martin and Morgan, along with Kunett, engaged in the creative process. Clementine Paddleford recorded Morgan's telling of events for the *New York Herald Tribune* in 1948:

> "We three were quaffing a slug, nibbling an hors d'oeuvre and shoving toward inventive genius." The three of them played around with their respective ingredients until they settled on a hefty shot of vodka, a few ounces of ginger beer and a squeeze of lime in a mug filled with ice. Morgan continued, "It was good. It lifted the spirit to adventure. Four or five later, the mixture was christened the Moscow Mule—and for a number of obvious reasons." (Miller and Brown, *Spirituous Journey: A History of Drink*)

Russian-style vodka combined with the likely "kick in the head" effects of consuming too much of the drink—the name works. The Moscow Mule was originally served in a signature copper mug embellished with a kicking mule. If your home bar is short a set of mule mugs, no worries. I am just as likely to reach for a Highball glass; it is delicious either way.

In keeping with the original's theme, I lean toward a bold, spicy, rye or mixed-grain vodka. Pairing it with the bold spicy tones of ginger beer prevents the spirit from disappearing from view.

2 oz (60 ml) vodka

½ oz (15 ml) freshly squeezed lime juice

4 oz (120 ml) chilled ginger beer

Fill a copper Moscow Mule mug with cracked ice, add vodka, lime juice, and ginger beer. Stir and garnish with spent lime shell.

MUDSLIDE

After one sip, you'll be sliding into a happy place.

RUM SHOP RYAN

Picture this divine constellation: summer afternoon sunshine, barely a breeze as you settle into a poolside repose, warmth that hugs and immobilizes you—it all says Mudslide to me. This decadent, frosty tipple is familiar to thousands of cruise vacationers lucky enough to visit Rum Point, the beautiful stretch of beach on Grand Cayman. There you will find the Wreck Bar and this legendary treat. At 400 Mudslides served daily, "popular" is an understatement.

Fans of the traditional White Russian will see a strong resemblance here. Though generally I'm not a fan of blender drinks, tending instead toward those shaken or stirred, the Mudslide delivers. I have grown fond of it for this was a favorite of my cousin Helen David. It became our frequently revisited "official" drink of a special celebration—when Helen and I joined a group of loyal customers from her bar, the Brass Rail, in Cancun, Mexico. The adventure left such an impression that Helen had me plug in the blender and crank out Mudslides whenever entertaining our Cancun travel compadres.

Part dessert, part libation, this is one decadent summer treat. Look for a vodka with a creamy mouth feel and notes of vanilla, toffee, cocoa, and/or nuts, which are characteristic of vodkas made from corn or potato.

2 oz (60 ml) vodka
1 oz (30 ml) Kahlúa
1 oz (30 ml) Baileys Irish Cream
1 oz (30 ml) heavy cream

Add the vodka, Kahlúa, Baileys, and heavy cream to your blender. Add 1 ½ cups cracked ice; blend until creamy. Strain into a chilled Hurricane glass. Top with chocolate nips and serve with a straw.

PINK SANGRIA

Your life will be defined by what you give, not what you take.

HELEN DAVID

I love Sangria. I love to make it, I love to serve it, and I love to drink it. Easy, festive, and fun, it is perfect for large gatherings and preferably should be made the night before to offer the host time for other preparations.

Introduced in the United States during the 1964 World's Fair in New York, Sangria has been enjoyed in Spain for hundreds of years. The traditional version is made with Rioja (red wine), fresh fruits, fruit juices, brandy, and soda. I fell in love with Pink Sangria when I was exploring the beautiful rosé wines of Provence in the South of France. This rendering is the signature drink of the Helen David Memorial Bartenders Relief Fund, supporting bartenders who have been affected by breast cancer.

One of Sangria's best features is that it can easily be adapted—every foodie's dream. The only rules for me are to include a selection of seasonal fruits and to consider the foods being served with the drink before committing to a white, red, or rosé wine base. As with any successful recipe, keep in mind that your final result is ultimately dependent on the quality of the ingredients you use—a philosophy ingrained in me by Helen David, my cousin and most prominent mentor. If you wouldn't drink the wine or vodka on its own, or eat the fresh fruit from the market, it does not belong in your drink.

I stay with an approachable, light, floral, fruit-forward vodka—generally characteristic of a New-World style made of wheat, mixed grain, or even quinoa. This recipe serves eight people—or four people, two each!

1 bottle (750 ml) rosé

8 oz (240 ml) vodka

4 oz (120 ml) Cointreau

4 oz (120 ml) simple syrup (see page 112)

4 oz (120 ml) pomegranate juice

8 oz (240 ml) freshly squeezed lemon juice

4 oz (120 ml) white grape juice

2 lemons, cut into thin quarters

6 strawberries, sliced

1 apple, sliced

1 cup (240 ml) red and green grapes, sliced

1 small orange, cut into thin quarters

Chilled 7-UP as needed

Place all ingredients (excluding the 7-UP) into a large glass container, cover, and refrigerate overnight. When ready to serve, pour into an ice-filled pitcher until two-thirds full. Add fresh sliced fruits and top with 7-UP; stir gently to mix. Serve in ice and fruit-filled wine glasses.

RUBY

True happiness consists not in the multitude of friends, but in their worth and choice.

SAMUEL JOHNSTON

The memories most enduring for me are those that involve great friends, great company, and great food and drink. Steak pomme frites with friends at Bouchon, Irish Coffee for two at the Buena Vista, hot dogs with my brother at the Coney Island Diner, and above all the Friday night pickerel fish fry at the Krystal Bar.

One such memory was an evening with friends Mike and Maggie in a wonderful little French bistro in Chicago for a night of lively conversation, delicious dinner, well-crafted drinks, and a lot of fun. I am always interested in trying new cocktails, though admittedly critical on the topic, so I was pleasantly surprised when the bartender offered us something new—a clever mix of vodka, elderflower liquor, and grapefruit juice. A great time with a great drink that has stood out in my mind ever since; the Ruby is a close rendering of that night's tipple. Aperol with its citrus and slightly bitter elements combined with elderflower's sweeter floral notes blend remarkably—quite different in profile but made better in combination.

The Ruby is a fruit-centric drink. I use a vodka with a good structure, rich mouth feel, and diverse fruit favors. A potato base is ideal to help soften and moderate the drink's bitter and floral elements.

1 ½ oz (45 ml) vodka

½ oz (15 ml) Aperol

¾ oz (22.5 ml) St-Germain Elderflower liqueur

¾ oz (22.5 ml) freshly squeezed Ruby Red grapefruit juice

¾ oz (22.5 ml) freshly squeezed lemon juice

1 tbsp (15 ml) egg white

In a mixing glass add vodka, Aperol, St-Germain, fresh grapefruit and lemon juices, and egg whites; shake with ice until well mixed and egg whites are emulsified. Strain into a chilled champagne saucer. Garnish with the oils of a grapefruit twist.

SGROPPINO

The trouble with eating Italian food is that five or six days later you're hungry again.

GEORGE MILLER

Drink or dessert? Call it an intermezzo—the perfect balance of acidity and sweetness to cleanse the palate during the course of a meal, typically in anticipation of the main course. This version was inspired by Mario Batali, whose hand and whisk conjured my first Sgroppino; this was back in 1993 when I was fortunate enough to work in Pó, his first New York establishment. I've been enjoying versions of this concoction ever since. Visitors to Italy may already be familiar with one variation or another.

Industry guru David Wondrich, in his book *Esquire Drinks: An Opinionated & Irreverent Guide to Drinking*, traces the Sgroppino to the northeast corner of Italy. He explains that one is served a Sgroppino, which loosely translates to the "little un-knotter," after a rich, heavy meal. I welcome this drink before, during, after, or entirely without a meal. It is simply delicious, refreshing, and super easy to throw together—a genuine crowd-pleaser, prime for interpretation. Switch out the fruity sorbet theme, and you're off to the races.

For this drink, I like an Old-World, creamy potato- or wheat-based vodka with a rich mouth feel—both kinds will help moderate the tart, citrus elements.

1 pint (480 ml) lemon sorbet, slightly softened

4 oz (120 ml) vodka, from the freezer

1 oz (30 ml) limoncello, from the freezer

8 oz (240 ml) chilled Prosecco

In a mixing bowl whisk together sorbet, vodka, and limoncello until smooth. Add chilled Prosecco and stir to blend. Transfer to a pitcher and serve in well-chilled champagne flutes. Garnish with fresh lemon zest, optional. Serves four.

VESPER

"A dry Martini," Bond said. *"One. In a deep champagne goblet."*

"Oui, monsieur."

"Just a moment. Three measures of Gordon's one of vodka, half a measure of Kina Lillet. Shake it very well until it's ice-cold, then add a large thin slice of lemon peel. Got it?"

"Certainly, monsieur." The barman seemed pleased with the idea.

IAN FLEMING, *CASINO ROYALE*

Injected into popular culture by Ian Fleming's 1953 novel *Casino Royale*, the Vesper pays homage to Vesper Lynd, James Bond's sultry yet ill-fated love interest. Far from delicate, this drink was purportedly invented and enjoyed by Fleming and his adventurous friend Ivar Bryce, whose copy of the novel is inscribed, "For Ivar, who mixed the first Vesper and said the good word."

The Vesper is a truly daring cocktail, featuring both gin and vodka with Lillet—a resplendent vermouth stand-in. Many people are enthralled by Bond's flawless, suave, yet rugged persona, but what an intriguing character Fleming must have been. After all, he named a beguiling woman of less-than-lily-white repute after a cocktail named for the sixth canonical hour—specifically, evening prayers. Fleming's cocktail party was definitely one I would have wanted to attend.

Feeling a little cautious? Consider this drink a wonderful bridge between gin and vodka.

Bond would approve of using an Old-World, grain-based vodka; I always use something bold, spicy, and assertive. Maybe you can go all in and venture into the world of the 100 proof distillate.

1 oz (30 ml) vodka

2 oz (60 ml) gin

½ oz (15 ml) Lillet Blanc

In an ice-filled mixing glass add vodka, gin, and Lillet Blanc; shake until ice cold. Strain into a chilled cocktail glass and garnish with a large, thin slice of lemon peel.

THE WIZARD

In a way, we are magicians. We are alchemists, sorcerers and wizards. We are a very strange bunch. But there is great fun in being a wizard.

BILLY JOEL

Much contributes to a bartender's signature style—principally, years of practice combined with influence and inspiration from others along the way. It's no secret, I tend toward fresh ingredients, generally fruit and citrus forward, and sometimes whimsical recipes. I always prefer to keep it fun. Striving for balance while honoring a featured spirit is also essential. As my profession continues to evolve, countless more trinkets and techniques will appear. Yet, staying true to classic methods rarely disappoints. For me, the two biggest rewards are succeeding in partnership with simplicity and seeing others make and enjoy my drinks.

Admittedly my repertoire has its fair share of complex recipes, yet I maintain that not every new creation needs fuss and frills—less really can be more. One way to maintain simplicity without sacrificing creativity is with a spirits-only cocktail, such as Manhattan, Martini, Negroni, Rob Roy, Just for Mary—all sophisticated, timeless cocktails made with relative ease. Along similar lines, the Wizard is a simple combination of spirits featuring lesser-known ingredients—Cinzano Bianco and Yellow Chartreuse. The results are far from simple, no bar-wizardry required.

The Wizard benefits from a robust vodka with lots of character. I look to the Old-World vodkas of Eastern Europe, those made from rye or potato.

2 oz (60 ml) vodka

1 oz (30 ml) Cinzano Bianco

½ oz (15 ml) Yellow Chartreuse

2 dashes orange bitters

In an ice-filled mixing glass add vodka, Cinzano Bianco, Yellow Chartreuse, and orange bitters; stir until well chilled. Strain into a chilled, small (Nick and Nora–sized) cocktail glass. Garnish with a thin slice of orange peel.

SUBRECIPES

SIMPLE SYRUP

Simple syrup is one of the essential cocktail staples—the best way to add the pure sweetness of sugar to a drink recipe.

1 cup (480 ml) water

1 cup (480 ml) sugar

Dissolve sugar into boiled water. Allow mixture to cool and store in a clean bottle in your refrigerator for up to a month.

GINGER SYRUP

2 cups (480 ml) water

2 cups (480 ml) sugar

1 cup (240 m) thinly sliced, peeled fresh ginger

Bring water and ginger to a boil in a 1 ½-quart saucepan, add sugar, and stir until dissolved. Return to a boil, reduce heat, and simmer uncovered for 20 minutes, stirring occasionally. Remove from heat and allow to steep until it reaches room temperature. Pour syrup through a sieve into a bowl, discarding the ginger. Bottle and refrigerate for up to one month.

APPLE-CUCUMBER PURÉE

1 lb (454 g) Granny Smith apples, unpeeled, cored, and quartered

1 oz (30 ml) fresh lemon juice

½ cup (120 ml) water

¾ cup (230 ml) sugar

1 English cucumber, peeled and diced

Place the apples, lemon juice, water, and sugar in a medium saucepan over high heat. Bring to a boil then reduce to a simmer. Cook the apples until they are extremely soft (about 20 minutes). Remove from the heat and allow to cool to room temperature.

Once the apples have cooled, transfer them to a blender with the diced cucumber and blend until the mixture becomes a smooth purée. Strain the mixture through a fine sieve. Refrigerate in a sealed container for up to three days.

FRESH GREEN APPLE PURÉE

2 lbs (1 kg) Granny Smith apples, unpeeled, cored, and quartered

2 oz (60 ml) freshly squeezed lemon juice

1 cup (240 ml) water

1 ½ cups (360 ml) sugar

Fraction of a drop of liquid green food coloring (apply with a toothpick)

Place the apples, lemon juice, water, and sugar in a medium saucepan over high heat. Bring to a boil, then reduce to a simmer. Cook the apples until they are extremely soft—about 20 minutes. Remove from the heat and allow to cool to room temperature.

Once the apples have cooled, transfer them to a blender and blend until the mixture becomes a smooth purée. Strain the mixture through a fine sieve. If you are seeking an apple green hue, mix in a tiny amount (literally, a drip from the end of a toothpick) of green food coloring. Place in a sealed container and refrigerate for up to three days.

PINEAPPLE-VANILLA BEAN INFUSION

2 ripe, fresh pineapples, peeled, cored, and sliced

2 whole vanilla beans, sliced to release seeds

2 bottles (1.5 liters) vodka

In a clean, wide-mouth jar place sliced pineapple and vanilla beans, and cover completely with

vodka. Replace lid and store in a cool, dark place for three to five days. Strain through a large, fine sieve to remove the fruit. Bottle and refrigerate.

Basic rules for foolproof infusions, as outlined in *The Modern Mixologist*:

1. Secure an infusion jar with a wide mouth and a tight-fitting lid.

2. Select very ripe, fresh fruits as they provide the best flavor and natural sweetness.

3. Rinse all fruits and vegetables before infusing.

4. Place fruits, vegetables, herbs, and spice in a clean infusion jar.

5. Select a neutral vodka, generally one that has New-World character and is grain based—either corn, wheat, or mixed grain.

6. Completely cover the infusion with vodka. (Reserve the empty vodka bottles, into which you can transfer the finished infusion.)

7. Store the infusion jar in a cool, dark place.

8. Most fruits and vegetables infuse in three to five days but the best method for determining whether your brew is finished is to taste it.

9. Stir the infusion daily and taste it to monitor.

10. When finished, strain the vodka through a fine sieve. Pour into the empty vodka bottles and refrigerate until ready to consume.

FLAVORED WATERS

To prepare flavored waters, start with the freshest vegetables and the ripest fruits of the season. Wash them well and prepare them as directed in each recipe. Combine the ingredients in a glass jar, preferably one with a wide lid, and add water. (Use distilled, reverse osmosis, or bottled water to avoid any unpleasant chemical or fluoride flavors.) Leave in the refrigerator overnight. Strain through a fine-mesh strainer and refrigerate until ready to use.

MANDARIN HIBISCUS WATER

4 to 5 small mandarins, quartered (or 3 clementines)

¼ cup (60 ml) dried hibiscus flowers (see Resources)

2 star anise

4 cups (1 liter) water

FRESH CUCUMBER RED BELL PEPPER WATER

2 cups (480 ml) diced cucumber

1 cup (240 ml) sliced red bell pepper

1 tsp (5 ml) black peppercorns

1 small bunch cilantro

4 cups (1 liter) water

FRESH BERRY WATER

1 ½ cup (360 ml) fresh and ripe mixed berries (raspberries, strawberries, blackberries)

1 long piece lemon zest (Peel a lemon like you would an apple, and avoid as much white pith as possible.)

4 cups (1 liter) water

PINEAPPLE CILANTRO WATER

2 cups (480 ml) diced, peeled very ripe pineapple

6 sprigs cilantro

1 tsp (5 ml) chopped, peeled fresh ginger

4 cups (1 liter) water

CAVIAR: VODKA'S BEDFELLOW

Although I came to know vodka well, I found caviar to be a bit of a mystery. I've always enjoyed it as a singular treat when given the opportunity—mostly when I've been invited to an exclusive or glamorous soirée. Mounded and featured in an ice-filled centerpiece accompanied by delicate blini pancakes, crème fraîche, and finely diced bits of egg and onion, caviar conveys a certain tone of prestige and elegance. Though not to everyone's taste, it is briny, rich, delicious—and quite frankly alluring.

Vodka with Caviar

For centuries, vodka and caviar has had an intriguing, luxurious connection—a consequence perhaps of the royal attentions garnered by both. The pair also is ingrained in Russian culture, as Begg confirms: "the combination is as evocative of the country as sable fur, metrushka dolls and the Bolshoi Ballet." A familiar indulgence among Russians and others in the Eastern European region, vodka and caviar to this day are reliably served side by side during any happy occasion—birthday, wedding, holiday, family gathering, and so on.

Faith and Wisniewski point out that caviar has "robust and intense flavors that vodka can balance." My culinary sensibilities agree—there are complementary elements involved.

Russian Tradition

Caviar's history is rich—reaching across continents, traversing millennia. A delicacy once reserved for royalty—notably the Tsars of Russia—and adopted as an indulgence among early European aristocracy, caviar remains a symbol of the passion and appreciation for the finer things. Inga Saffron in *Caviar: The Strange History and Uncertain Future of the World's Most Coveted Delicacy* explains that caviar's affluent and coveted tradition may come from its "intense flavor, high price and most of all its renowned scarcity." Nichola Fletcher in *Caviar: A Global History* passionately recalls her first encounter with this delicacy: "As soon as

I licked my first mouthful off the toast I was captivated, I wanted to feel again and again that sensation of soft beads melting into something primeval yet tantalizingly elusive; the absolute essence of ancient seas."

Delicious and irresistible to some, expensive to many, mysterious to most. What is it exactly?

Caviar is the delicate roe (unfertilized eggs) harvested from sturgeon, rinsed in water, and lightly salted. Roe from the Sevruga, Ossetra, and Beluga sturgeons remain the most valuable and the most expensive because they are challenging to procure. In the old days or traditional sense, if the roe did not come from the wild sturgeon native to the Caspian or Black seas, it was not considered *true* caviar.

The three most familiar varieties of roe are Sevruga, Ossetra, and Beluga. Each type has a characteristic size, color, texture, and flavor; environmental conditions—such as water quality and the sturgeon's food source—further influence its harvest. Caviar purveyors identify and designate prized "crops," and this designation commands exceptional prices. The following is a general description of each type:

- Sevruga is the smallest of the sturgeon family, weighing between 50 and 60 pounds. This fish produces small, grayish roe that has relatively salty and buttery flavor. The Sevruga roe is the most affordable of the elite three.

- Ossetra sturgeon can weigh over 400 pounds and produces roe whose color ranges from light to dark brown with hints of gold. The Ossetra roe delivers a distinctive nutty flavor.

- Beluga sturgeon is not only the rarest but also the largest at over 2,000 pounds at maturity, which takes as long as 20 years. Its roe is large in size, has fine skin, and has a rich and creamy texture. It commands the highest price of the three varieties—largely because of its scarcity.

Due to overharvesting, the wild sturgeon traditionally hailed as the producer of "true" caviar has become increasingly rare and hence progressively expensive. Nearing extinction, the wild sturgeon is no longer harvested from Russian waters, as the country implemented a self-imposed ban in 1998. Furthermore, in 2005 the Beluga sturgeon was added to the endangered species list by the United States and neither it nor its caviar can be imported and sold within the US borders.

The caviar industry's move toward farm-raised sturgeon has, to a large degree, kept fine caviar from disappearing altogether. Sturgeon aquaculture is a highly specialized craft mastered by only a few but has great reward potential, turning out hundreds of millions of dollars in profit annually. Iran, Russia, Italy, and several countries along the Persian Gulf are among the largest producers, but despite healthy production prices remain high. A single mature sturgeon (maturity takes anywhere from 8 to 20 years) can be worth more than $15,000.

Words from a Caviar Expert

Intrigued and inspired to look deeper into caviar, I paid a visit to Alexandre Petrossian, esteemed member of the Petrossian caviar dynasty and director of its New York retail operations. What I learned was fodder for a text of its own.

Two brothers—Melkoum and Mouchegh Petrossian—are credited for introducing caviar to 1920s Paris; a move that ultimately influenced its luxury status to the rest of the West. The City of Lights provided a welcoming safe harbor to many Russian aristocrats and artists exiled by the Bolshevik Revolution, as well as to the Russian cultural and culinary traditions—caviar among them—these exiles brought along. The Petrossian brothers capitalized on Parisians' developing enthusiastic desire for Russian fineries. It was the embrace of the brother's fledgling venture by luxury hotelier César Ritz that elevated caviar to a broadly ingrained symbol of fine dining for Parisians and beyond.

From Staple to Specialty in the United States

Sturgeon and its roe were once a common food source in the United States, used by early European colonists and Native Americans before that. Fletcher adds, "Although they clearly ate a lot of sturgeon, it cannot be said that the settlers relished it as their Russian counterparts would have." Sturgeon was not a status symbol but rather a staple. Well into the 19th century, sturgeon was consumed in great quantities because it was widely available and reasonably priced. Its salty roe was served in local saloons to keep patrons thirsty.

According to Fletcher, two immigrants to the United States—one Russian, one German—were pivotal to shifting the culture's use of sturgeon from staple to specialty. The legend was that in the 1840s, a Russian émigré purchased sturgeon from a fisherman along the Delaware River for "the good price of a dollar." He prepared the roe in such fine manner that its quality was high enough to export to the Russian and European markets, which flourished in the decades to come. The German immigrant in the story was Bendix Blohm, a fisherman by trade. He essentially fashioned the earliest version of US sturgeon aquaculture. Fueled by the success of his initial sturgeon export, Blohm enlisted a group of six fishermen to raise the fish in ponds. Once matured, the sturgeon were harvested and their roe converted into caviar.

Profitable ideas being infectious, the caviar industry underwent a meteoric ascent in the United States. Bodies of water swollen with the "bunionesque" creatures became fishermens' pot of gold, and within 50 years "the price for a sturgeon swollen with eggs had jumped to thirty dollars," according to Saffron.

By the end of the 19th century, caviar had gained attention as a lucrative export. Simultaneously the idea of caviar as a delicacy began to catch on, and this new appreciation radically increased demands in the United States. Shortly after, severely depleted US supplies and the developing preference and prestige associated with caviar from Russia meant that imports eventually trounced exports. By the end of the 20th century, the United States surpassed Russia in caviar consumption, coveting above all the roe from the three most-prized sturgeons—Sevruga, Ossetra, and Beluga.

Words from a Caviar Expert

Asked if Beluga would be farm-raised in the United States, caviar expert Alexandre Petrossian responded that it could happen—but not "for at least the next ten years. It takes a female Ossetra between five and seven years to mature enough to produce eggs; the Beluga takes much longer, up to 20 years." He stresses, however, that there are a variety of high-quality, farm-raised, imported and domestic caviars worth our consideration today.

Alexandre promotes farm-raised Kaluga, known for roe with a mellow, rich, buttery characteristic, as a close alternative to Beluga. Still for many, it may be an expense beyond reach. Instead, there is the American sturgeon, an affordable quality caviar ideal for those with a limited budget or those venturing into caviar for the first time. It is, Alexandre says, "approachable and delicious with a balanced taste and texture." So what's not to like?

Many roe varieties are also available from other fish, including whitefish, North Atlantic salmon, lumpfish, paddlefish, carp, and cod. These are considerably less expensive, and some are considered by experts as respectable while others are best for cooking. But do note this: If the designation "caviar" is used, the label of the roe should state the fish species—consider it a denomination of origin. As with most things, you get what you pay for. If the price seems too good to be true, proceed with caution, as your caviar experience may be an unpleasant one.

How to Serve

Again, serving caviar is purely a matter of choice. Often caviar is served amid a circle of sides—blini, diced red onion, diced hard-boiled egg yolk and white, and crème fraîche. But let me invoke my philosophy about drinking vodka on its own and apply that to tasting caviar without accompaniment. Caviar is such a rare and extraordinary indulgence. Why mask top-shelf caviar's full flavor expression by introducing any distractions?

In truth, few (if any) experts would recommend eating caviar with anything else. The Petrossian company also advocates the "naked" approach, stating on its website, "Serve fresh premium quality caviar in its original state of perfection… save these garnishes for inferior grades of caviar." That said, caviar that is less than premium grade or with strong flavor elements can benefit from the addition of freshly made blini (page 122) and a dollop of crème fraîche.

Here are some practical tips for storing and serving caviar:

Storing: Caviar is extremely perishable and must remain refrigerated until ready to consume. It can be stored for up to four weeks—unopened—in the refrigerator; once opened, it should be enjoyed immediately. Word of caution: Never freeze caviar; its delicate nature will not survive or be palatable.

Serving: Ideally it should be removed from the refrigerator only 15 minutes before serving. Too cold, and some of its full flavor spectrum is lost. A purely visual reward, the caviar tin set in a crystal or shiny silver bowl filled with crushed ice is a simple yet striking presentation. If you want to provide traditional or nontraditional sides, arrange them around the caviar bowl. You and your guests are about to enjoy a true delicacy and a rare extravagance, so find your style and make a show of it!

Quality caviar is best delivered straight from the tin to the mouth. Serve using only a mother-of-pearl or nonmetallic spoon, as metal adversely affects the caviar's flavor. Some enthusiasts prefer to spoon a mound onto the back of their hand—roughly between the thumb and index finger—introducing it into the mouth from there. A strange-looking ritual, this approach injects a tiny amount of warmth from the skin into the caviar and removes a bit of the caviar's chill such that "flavor and perfume emerge," rendering a more dynamic reveal on the palate, according to Fletcher. I have to admit, this really clicked for me, as the same principle applies to vodka. As I explained in an early chapter, vodka's full expression isn't accessible at its coldest; only as it begins to warm will the palate have access to its full flavor potential.

When all is said and done I prefer my caviar as I do my vodka—neat. Having tried all the aforementioned foodie accessories, give me caviar without garnish and a glass of premium freezer-cold vodka. A glorious tableau: me, ice-cold vodka, a tin of caviar, a mother-of-pearl spoon. The only improvement may be to double it all for sharing, making a simple but elegant celebration for two.

A Novice Vodka–Caviar Pairing Experiment

The urge to explore the flavor interplay between the historical bedfellows vodka and caviar proved irresistible for Mary and me. So during one of our writing weekends, we set out on a late-night "academic" exercise of tasting caviar and vodka. We found ourselves ordering away, blissfully ignoring the financial outlay, and decided on three caviars and three vodkas—we ordered a mix of caviar and vodka that spanned domestic and imported Russian. The vodkas were served chilled and neat, while the caviar tins arrived cold on a bed of ice along with the usual sides.

Our first approach was decidedly novice-level and admittedly unorganized. We did, however, find clear vodka–caviar pairing winners as well as pairings that clashed on our palate. We opt not to name the brands we tested, but here's the result of our first experiment: Domestic corn-based vodka paired best with the domestic farm-raised paddlefish caviar, while the Russian wheat vodka was superb with the imported Russian Ossetra caviar.

Intrigued yet unconvinced by these results, we pushed on to a second experiment. This time, we were more organized—thanks to donations made by Petrossian caviar and the participation of our photography team at Tim Turner Studios. We decadently tasted three different caviars and six vodkas that span the ingredient and character scope. The caviar selection included Alverta President (domestic), Royal Transmontanous (domestic), and Royal Ossetra (imported); each caviar has unique qualities.

Our testing group included five people, and we each leaned toward a favorite caviar but unanimously agreed that the Royal Ossetra edged out the others. For

FRESH BLINI

MAKES APPROXIMATELY 40 TO 45

1 cup (240 ml) self-rising flour, less one tbsp

1 tbsp (15 ml) Bob's Red Mill buckwheat flour

Pinch salt

1 egg yolk

2 egg whites

½ cup (120 ml) milk, room temperature

½ cup (120 ml) soda water, room temperature

4 tbsp (60 ml) melted butter

PREPARATION

Combine in a bowl both self-rising and buckwheat flours and salt. Mix together soda water, milk, melted butter, and egg yolk. Beat egg whites until fluffy but not dry. Blend milk mixture into flour mixture until most of the lumps are gone, then fold in the egg whites. Try one blini to assess if batter thickness is desirable. Adjust thickness, as needed, with equal parts milk and soda water. Your goal is to produce a thin, golden brown pancake—one teaspoon-size portion of batter should make one blini roughly 1 ½ inch in diameter and 2-3 millimeters thick.

COOKING

Heat a nonstick skillet or griddle over medium heat. If needed, lightly coat it with butter or cooking spray (if butter burns quickly, reduce heat). Spoon approximately 1 teaspoon batter into skillet. Cook until golden brown, turning only once. Cooking time is approximately 1 ½ minutes total—once temperature is correct, batter should take about 30 seconds on one side and 45 seconds on the other. Once batter and cooking surface temperature are adjusted as needed, make up to four at a time. Best consumed the same day.

the pairing, most of us declared the Royal Transmontanous as the best match for a potato vodka, the Alverta President as complementary with both mixed-grain and corn vodkas, and the Ossetra as the most suitable for rye vodka.

The pairings that were favorable were in no way related to the vodka's or caviar's country of origin. What did account for the matches we found were the complementary elements between vodka and caviar. For example, spicy rye vodka elevated creamy, buttery, nutty caviar, while a creamy feminine vodka balanced the caviar with strong, briny flavors.

Delightful Conclusions

Again, our pairing approach was decidedly novice, even self-indulgent, but truly enlightening. In the end we shared a couple of wonderful, memorable evenings; learned some lessons; and fueled our excitement about vodka.

First, the notion that not all vodkas are the same was absolutely reinforced. Each one interplayed uniquely with the flavor of one caviar to the next. With some effort and a level of interest, vodka's flavor and character can be taken further— paired with caviar, mixed in a cocktail, or chilled in the freezer.

Second, the bond between vodka and caviar is alive and well—just as the Russian elite had known and everyone else had coveted many centuries ago. Enjoyed together, vodka and caviar are irresistible. Enjoyed with good company, vodka and caviar are magical.

See for yourself what we mean. If pairing vodka with caviar is within your budget, go for it. I can almost guarantee a wonderful experience.

CHAPTER FIVE

Taste and Tasteability

ONE OF THE HONORS BESTOWED ON EXPERIENCED INDUSTRY FOLK IS A SEAT on an internationally recognized spirits-tasting panel. The San Francisco World Spirits Competition, the Ultimate Spirits Challenge in New York, the International Spirits Challenge in London, and the International Wine & Spirits Competition are examples and are strictly for the "big" boys and girls of our field. Panel members take on the task of evaluating, comparing, and competitively scoring entries, which sometimes total upwards of a thousand. A charge taken quite seriously, as the published final scores have a considerable commercial impact.

Two common denominators exist in professional tastings:

1. All tasting is blind. There are absolutely no visual cues that define the brand in the glass.

2. Spirits are always tasted at room temperature.

Gathered in small groups of four (typically), the panel is presented each spirit to be tasted—neat, at room temperature, in a glass marked only by a number. The only variables, and yes they are considerable, are the taster's experience and palate. And I would add fatigue, but more on that later.

Every vodka included in this book was tasted in exactly the same fashion used at the professional tasting events (for the full review, see the section Exploring 58

Vodkas—One Brand at a Time later in this chapter). I admit that it takes practice to exercise and fine-tune those tasting muscles, but anyone with a sense of smell and taste can find the experience revealing. In truth, if we take the pressure out of tasting, it becomes an enjoyable practice.

HOST A VODKA TASTING

Hosting a tasting for like-minded spirit enthusiasts is a great way to introduce and explore vodka's intricacies. A chance to interact and learn from each other's experience and preferences and to ultimately introduce a tangible appreciation for vodka's nuances and versatility. It's not only educational but also a lot of fun. As the host, you forfeit a portion of the overall experience; after all, someone has to select, purchase, and pour the vodkas and secure the blind aspect for the panel. Regardless, it will be an excellent time for all, so give it a go!

HOW TO PREPARE

A fair amount of preparation is necessary to host a successful tasting party. Here's a step-by-step guide.

Select the vodkas for tasting: Choose a reasonable number of vodkas to allow enough variation in style and character and to allow dynamic comparisons. But don't provide too many as that will fatigue or overpower the palate's ability to discern. Six is fine if you are bent on including each of the base ingredient categories, but you may want to stop there. Three is more realistic and manageable. Include both bargain and premium brands, and select from different categories— rye, wheat, potato, mixed grain, and corn or something less traditional such as rice, grape, whey, maple sap, and quinoa.

Keep the vodka selection a secret: As humans, we instinctively give and receive impressions based on visual cues. Without the visual to guide us, our preferences are often surprising. Vodka bottles and packaging are distinctly shaped and familiar, so keep them out of view to maintain the blind nature of the tasting. Jot down a tasting key so that when all is said and done you can easily reveal each brand to the participants.

Plan for enough time: Approximately 10 to 15 minutes are needed for each vodka—nosing, tasting, discerning, taking notes, cleansing the palate, and discussing. For three vodkas give at least 30 to 45 minutes—and that is moving along at a healthy pace. To ensure a nice, relaxed pace, allow the tasters enough time.

Decide the size of the tasting party: How many people to invite is largely dependent on the space available. Keep in mind, everyone should gather around one table for ease of interaction and exchange of impressions. Six would make the party lively, but fewer is fine if space is limited.

Provide appropriate glassware: Tasting glasses are stemmed, tulip shaped, relatively narrow in diameter, and range from 3 to 4 ounces in capacity. The closest comparison is a sherry glass, which is similar to a miniature version of a traditional wine glass. You can also make do with very small white wine glasses. Avoid glassware with a wide-diameter bowl or opening. As for quantity, multiply the number of brands included by the number of participants. For example, if you have 3 vodkas and 6 guests, you will need 18 glasses.

Before using wash each glass in warm water and decent dishwashing liquid, then rinse thoroughly to remove any trace of soap residue or odor. Once dried, buff with a clean cotton cloth—one devoid of odor from scented laundry detergents or dryer sheets. Think neutral.

Number each bottle and tasting flight: Arrange your brands in any order you choose. Use a Sharpie marker to number each bottle; mark also the base of each set of glasses—a "flight." For a party of six tasting three brands, you'll end up with six flights—six sets of glasses, numbered 1 through 3. Despite its reputation as a permanent marker, Sharpie ink easily washes off glassware with dishwashing liquid and a sponge. (Notice any spills when pouring. Vodka behaves like a solvent, so you may need to renumber.)

Pour and cover: Pour approximately 1½ ounces of vodka into its corresponding numbered glasses. Once poured, cover each glass immediately with a paper muffin or cupcake liner—placed upside down atop the opening. Without covering, the vodka's subtle aromatics will be quickly lost, so it is important to cover immediately after pouring. The covering is not removed again until you are ready to pick up the glass and go to work.

Set up the tasting area and table: For each participant, set a place with the following:

- **Tasting sheets and pencils or pens.** Findings can be handwritten on a blank sheet of paper, but making your own tasting sheet (page 129) helps organize the process. It is best to follow an order by which tasting is performed. For vodka I recommend this order: nose, palate, mouth feel, finish, character, overall impression or suggested uses. With a white spirit, color isn't a consideration. Professional tastings also include scoring—include it if you like.

  On the back of the tasting sheet, include descriptors (page 130) that tasters (especially novices) can draw from or refer to when documenting their findings. This is a list of terms for describing the vodka's aroma, flavor, and sensations one might discern.

- **Disposable spit cups.** Unfortunately there are no classy substitutes for this term—it is what it is. Each person needs his or her own spit cup, which should be plastic, disposable, opaque, and large. Nothing see-through, and nothing you will put back on the shelf to reuse.

- **Palate cleansers.** Provide everyone with still bottled water (no flavoring); neutral crackers such as water crackers (no salt or flavoring); and bite-sized slices of neutral, aroma-free cheese (domestic Muenster is ideal). Your palate retains memory after tasting, so taking a break

VODKA TASTING SHEET

Vodka Number: _____

Taster Name: _____

Nose: _____

_____

_____

Palate: _____

_____

_____

Mouth Feel: _____

_____

Finish: _____

_____

Character: _____

(1-3) neutral or subtle—more feminine; (4-6) balanced between neutral and robust; (7-10) robust or flavorful—more masculine

Overall Impression/Suggested Uses:

_____

_____

_____

_____

Score: _____

Between (1) No thanks, not for me and (10) Best ever, superior

VODKA TASTING DESCRIPTORS (Typical but not all inclusive)

HERBACEOUS—thyme, oregano, sage, tarragon, cilantro, basil

EARTH—chalk, clay, soil, flint, limestone, minerals, moss, mushroom

GRAIN—sweet corn mash, malty grains, cooked grains, yeast, toast, porridge, cooked cereal, biscuits, rice, corn, wheat, rye, bread dough, toasted grains

COOKING SPICE—black pepper, white pepper, anise, cardamom, caraway, coriander, sesame, cumin

BAKING SPICE—cinnamon, clove, ginger, licorice, nutmeg, vanilla, allspice

CITRUS—lemon, bitter orange, sweet orange, grapefruit, lime, kaffir lime, tangerine, mandarin, blood orange, Meyer lemon, kumquat

CITRUS PEEL—lemon, lime, orange, grapefruit

TREE FRUIT—apple, green apple, baked apple, pear, plum, fig, black cherry, nectarine, peach, apricot

MELON—watermelon, honeydew, Crenshaw

TROPICAL FRUIT—banana, guava, kiwi, mango, passion fruit, pineapple, papaya

BERRIES—strawberry, raspberry, currant, cassis, boysenberry, blackberry, blueberry

DRIED FRUIT—fig, prune, raisin, date, apricot

SMOKE/CHARCOAL

FLORAL—fresh-cut grass, hay, lavender, lily, orange blossom, rose, violets, lemon verbena, lilac, honeysuckle

NUTS—walnut, almond, hazelnut, cashew, pecan, toasted

ALCOHOL—soft, slight, balanced, pronounced, present, moderate, prominent, warm, hot, fiery

SWEET—toffee, butterscotch, caramel, cocoa, nougat, chocolate, crème brûlée, honey, molasses, sweet vanilla, maple syrup, cake batter, cream soda, marzipan, orange marmalade, bittersweet chocolate, confectioner's sugar, bread pudding, toffee, white chocolate, mocha

DAIRY—egg cream, sweet butter, cream, malted milk

MINT—camphor, eucalyptus, menthol, peppermint, spearmint

VEGETAL—boiled/mashed potato, radish, celery, green bell pepper, cucumber, fennel, asparagus

TEA—black, green, herbal, chamomile, hibiscus, mint

MOUTH FEEL

- Creamy
- Oily
- Acidic
- Dry, semi-dry
- Silky
- Luscious
- Satin
- Viscous
- Sticky, cloying
- Velvety
- Rich

- Clean
- Crisp
- Astringent
- Hot
- Cool
- Light, medium, or full bodied
- Acidity—light, medium, balanced, pronounced

FINISH

- Short, moderate, long, very long
- Lingering
- Memorable
- Pleasant
- Cool, warm, hot
- Finish with the memory of —sweetness, pepper, spice, toffee, bitterness, etc.

OVERALL CHARACTER

- Complex
- Balanced
- Layered
- Subdued
- Elegant
- Finessed
- Aggressive
- Masculine
- Robust
- Feminine
- Approachable

to cleanse the palate in between each round is essential to neutralize and prepare the senses. Time spent cleansing the palate also offers an excellent opportunity to make notes and share impressions.

HOW TO TASTE 101

Everything is now in place and ready to roll. Before you begin, a simple review of the tasting process will get all participants aligned and proceeding in a similar way. The goal is to encourage everyone to follow the course you set so that the brand comparisons—from start to finish—are done on an even or a level playing field.

Many of us have tried or been instructed in the process of wine tasting. Generally the principles of vodka and wine tasting are similar—with two key differences that I highly recommend for those taking on spirit tasting for the first time. This advice is aimed at preserving tasters' ability to get through the party with their nose and palate reasonably intact.

1. When evaluating the spirit's nose, do not smell or sniff the aroma through your nose (despite your instinct to do so); breathe in through your open mouth.

2. When tasting, do not swallow any of the sample (despite your instinct to do so).

Since most consumers are not schooled in or professionally acclimated to spirits tasting, erring on the side of protecting your senses is good advice—certainly more likely to facilitate a better overall experience. Remember, more than the case with other spirits, the variations in vodka from brand to brand are subtle, which is all the more reason to protect yourself when closely examining them. Following these recommendations is a matter of personal preference and experience. Many spirits and tasting professionals have different methods, and I myself use a slightly different approach to the one outlined here.

Tasting Order

That said, the order for tasting and evaluating any spirit are color, nose, palate, mouth feel, and finish. With vodka, add an evaluation of its character and exclude color. As you move through each component, jot down your impressions on the tasting sheet for the corresponding glass number.

Color: Very few vodkas—those not flavored—have color. Vodka and gin are the two spirits for which color can safely be left off the tasting sheet.

Nose: Always start with aroma. Remove the glass covering, and give the glass a few short swirls. Place your nose over the glass with your mouth open, and breathe in through your mouth and *not* your nose. Actively taking air in through your mouth stirs the air, allowing aromas to ride along this slight movement and passively enter the nose's olfactory centers. Try it several times, and in between give your glass a few swirls to aerate the vodka. It will continue to open up and express itself over the span of a few minutes. Think about what aromatics come through—grain, yeast, grass, citrus, floral, spice, vanilla, caramel, and so on. Pay attention to whether alcohol is prominent, balanced, or subtle on the nose.

Palate: Take in one-fourth to one-half ounce of vodka and swirl to completely coat the inside of your mouth, then discard into your spit cup. The palate is now primed for tasting. Take in another one-fourth to one-half ounce and swirl to completely coat your mouth. After a few seconds, discard into your spit cup. Before and after discarding, notice what flavors are revealed—some may be very prominent up front and others may develop over time, producing a range or flow of flavors from start to finish. Similar flavors previously found on the nose may dominate, or they can be quite different. Notice also the intensity of the alcohol in your mouth—soft, present, forward, prominent. Alcohol can balance and support flavors or be overly assertive, even unpleasant in a way that is disconnected or out of proportion with accompanying flavors.

Mouth feel: Before discarding, experience how the vodka feels in your mouth. Is it creamy, thin, viscous, silky, smooth, oily, clean? What is the acidity level—not the acid flavor but the sensation brought on by acidity? Acidity acts to balance sweetness and to add complexity. Its presence is desirable and lends structure to the spirit. In this context good acidity is present and stays with you, extending the flavor experience and finish. Think of the sensation that acidic food or drink elicits at the back of your mouth (often inducing saliva production)—specifically, where your molars meet your cheeks. Is acidity present, absent, slight, medium, prominent? At this point, take a deep, yoga-type abdominal breath. Does your palate cool down or remain hot?

Finish: Note the finish. How long do the flavors in your mouth persist, and what are you left with? The length can be short, moderate, long, very long, and so on. Are there any lingering flavors or impressions—heat, cool, spice, bitterness? After you took a deep, abdominal breath did you feel warm and pleasant in your chest or were you left with a hot alcohol afterglow? Does the flavor stop short and finish in your mouth, abruptly dropping off, or does it linger with alcohol heat and move downward?

Break to cleanse your palate: At this point, take a break to review your notes, compare them with your tasting mates, and cleanse your palate. Drink some water, and nibble on a cracker and a piece of cheese. Allow a bit of space before moving on to the next glass in your flight.

Character: The best way to sum up character is to think of all the elements—nose, palate, mouth feel, and finish—all together. Consider where the vodka falls along a continuum, using this range: (1–3) neutral or subtle—more feminine; (4–6) balanced between neutral and robust; (7–10) robust or flavorful—more masculine. Because tasting is purely a matter of preference, one range is not necessarily better than another. The value in discerning character is matching it with your drinking style. A big, bold, spicy, peppery, masculine vodka is a great

TYPICAL VODKA-TASTING FINDINGS BY BASE INGREDIENT

Nose

Wheat	Notes of vanilla and anise, citrus, white pepper, yeast, malty grains and bread dough
Rye	Big, black, spicy pepper notes; earthy and vegetal with hints of green bell pepper
Barley	Bright citrus, floral, and herbal character as well as spicy, nutty notes
Corn	Layers of grainy, cooked cereal, corn mash, sweet butter, and egg cream
Potato	Earthy, musty, fruit-cellar qualities; fresh-cut grass with mashed potato nuances

Palate

Wheat	Toffee sweetness with toasty grains (cereal), vanilla, caramel, and cream soda
Rye	Spicy pepper with a nutty, subtle sweetness; bittersweet chocolate
Barley	Roasted nuts, toffee, and vanilla; baked apple; white pepper spice
Corn	Buttery sweetness, corn-on-the-cob flavors, rich caramel, toffee, and a touch of cocoa and vanilla
Potato	Complex layers of rich cocoa, sweet vanilla, buttery mashed potatoes, and stewed dark tree fruits

Mouth Feel

Wheat	Softer with medium weight, good acidity, and a dry finish
Rye	Medium to full bodied that is rich upon entry but finishes dry with medium acidity
Barley	Light to medium bodied with a slight creaminess that finishes clean and crisp
Corn	Rich; silky; medium to full bodied with nice viscosity and light acidity with a semi-dry finish
Potato	Rich; luscious; full bodied; creamy with velvety viscosity that coats the palate

match for a Bloody Mary. While a feminine, approachable, citrus-forward vodka works better in a lighter, fruit-based cocktail. Like a Vodka Martini? Something balanced, full of flavor complexity, and with a long warm finish may become your go-to brand. The logic here is to zero in on your personal preference and apply it as you desire.

Last but not Least: The Reveal

With flights completed and notes compared and admired, it is time to pull out your tasting key and reveal the brands. I can almost guarantee surprises, and certainly there will be lots of fun. It's fascinating to discover how preferences usually align with price point and name brands (instead of lesser known) and how and whether the ingredient base is reflected in the findings.

Regardless of whether the participants came up with similar results or revealed marked but informative taste differences, consider your tasting a success! You've delivered a platform for developing vodka preferences and sparked renewed enthusiasm among those inclined toward a vodka drink every now and then.

EXPLORING 58 VODKAS—ONE BRAND AT A TIME

Rounding out this book is an individual profile for 58 vodka brands. Each one was tasted and evaluated by a panel of industry leaders. Why 58 vodkas? Admittedly that number is infinitely more than anyone should attempt to taste outside the professional tasting arena. The advantage in this number, however, is that it allows for a solid representation of a broad selection of brands that fall in all base ingredient categories—rye, wheat, potato, mixed grains, corn, and others.

Because it is extensive, the collection we feature here offers a good range for vodka character comparison, including regional exposure. Each brand is outlined according to the most influential and easily identifiable expression variables.

However, we present the brands not in order of rank but in terms of the vodka's source, character, strengths, and other qualities. Each brand profile provides enough information that is both useful and interesting without being overbearing. Here are the categories featured for each brand:

- **Style:** Old World or New World

- **Origin:** Country, region, or state of manufacture

- **Character score:**
  A scale of 1 to 10, ranging from subtle and feminine (1–3) through balanced (4–6) to bold and masculine (7–10)

- **Cost:** Based on 750 mL bottle size, retail price:
  - $ = 10–19 US dollars
  - $$ = 20–29
  - $$$ = 30–39
  - $$$$ = 40–49
  - $$$$$ = 50+

- **Material and techniques (ingredients and distillation/filtration process):**
  Including fermentable sugar and water source; pot (batch) or column (continuous) distillation

- **Tasting notes:** Nose, palate, mouth feel, finish, and character

- **Expert notes:** Tasting panel's commentary for suggested uses

Ultimately the profiles serve as a tool or reference for comparing, selecting, and enjoying vodka. Flip through them, and find a few brands that match your taste preference. Then, using the tasting guidelines we presented earlier, see how your own impressions stand up against those of our expert panel.

Note: All style, character scores, tasting notes, and expert notes in each profile are the opinions of the tasting panel members. All origin, cost, and material and techniques details are obtained directly from each brand or its industry representative, through written or verbal interview. In cases where specific details could not be obtained directly from the brand or representative, we used information that is publically accessible, including the brand's website. Any information that could not be secured by either method is marked "proprietary." Any brands not available in the United States at the time of this printing are marked thusly.

Our Panel of Tasting Experts

For this significant part of the book, we invited three professionals renowned for their knowledge of spirits and cocktails and their venerable ability to mix them. Our panel (which includes me) consists of Bridget Albert, Dale DeGroff, and Steve Olson. Each one serves as a professional taster in national and international spirits competitions. Each one has an arm's-length list of accomplishments, awards, and accolades. Each one I have had the honor and privilege to work alongside in one capacity or another, including judging spirits competitions.

BRIDGET ALBERT is a talented mixologist and educator currently based in Chicago. She developed her skills tending bar at the Bellagio in Las Vegas. She is the general secretary of the United States Bartenders Guild, a columnist for *Tasting Panel* magazine, and the coauthor of the book *Market-Fresh Mixology: Cocktails for Every Season*. In addition, Bridget is the director of mixology at Southern Wine & Spirits of Illinois and is the founder of the Academy of Spirits and Fine Service—a unique program for bartenders that covers the history of all spirits and pre-Prohibition cocktails. On a personal level, I know Bridget has a flawless reputation for professionalism in our industry, and she continues to inspire all she encounters to be the best they can be at their craft.

DALE DeGROFF, aka King Cocktail, is a renowned mixologist and educator known for pioneering and applying a gourmet approach to cocktails—in particular, reviving the great classics. His exceptional talent and techniques developed while tending bar at many legendary establishments. Author of *The Craft of the Cocktail* and *The Essential Cocktail*, Dale has been greatly influential in the mixology profession, setting in motion the cocktail explosion that continues to transform our industry today. One of his many accomplishments is that he is the founding president of the Museum of the American Cocktail, a nonprofit museum located in New Orleans that celebrates the rich, 200-year-old history of the cocktail. I met Dale in 1993 at the Rainbow Room, an encounter I consider pivotal in my drive to make the same footprint of dedication and professionalism as he consistently shows. It was Dale's recommendation to Steve Wynn that led to my accepting the challenge of master mixologist for the Bellagio in Las Vegas.

STEVE OLSON, aka Wine Geek, is dedicated to the "education and consultation of degustation for appreciation and celebration." Steve is widely recognized as an industry authority with a true gift for teaching. He also writes about wine, beer, spirits, sake, and virtually any other beverages under the sun. In March 2006 he, along with four partners (Dale DeGroff among them), opened Beverage Alcohol Resource, LLC, an independent organization whose mission is to educate, guide, and propagate the healthy, enlightened, and responsible use of beverage alcohol products. I first met Steve when I lived in San Francisco in the early 1990s. I was working as a bartender, and he was giving a gin educational seminar—the first such pursuit I had ever seen. Steve is the first person to open my eyes to the approaches and nuances of tasting spirits—skills I draw upon to this day.

LIST OF 58 VODKAS

RYE VODKAS

BELVEDERE

Style and Origin	Old World; Zyrardów, Poland
Character Score	7.5
Cost	$$$
Materials & Techniques	Dankowskie golden rye; purified water sourced from two proprietary artesian wells; four-times distilled in copper-lined column stills; filtered through charcoal and then cellulose
Tasting Notes	
Nose	Spicy; prominent ground black pepper; rye bread; vegetal—asparagus, green peppers; herbaceous; lemon peel; just the faintest hint of vanilla
Palate	Notes of white pepper, fennel, anise, cinnamon, and sweet vanilla; slight graininess; dark cooked fruits—plums and cherries, almost jam-like
Mouth Feel	Medium to full-bodied viscosity; slightly oily with beautiful acidity
Finish	Medium to long; warm and lingering with just a touch of toasted almonds
Expert Notes	"A vodka to drink straight"; "Anything savory; the Bloody Mary family"; "Tastes like rye bread"; "Pass the onions...Gibson time!"; "Big and robust where the alcohol is prominent but never overwhelming"; "Very balanced, well distilled"

CHOPIN—RYE

Style and Origin	Old World; Krzesk-Majatek, Poland
Character Score	7
Cost	$$$
Materials & Techniques	Golden rye from the Podlasie countryside; purified artesian well water; four-times distilled—once in a copper column and three times in rectification columns; five-times filtered—paper, then two candle filters, disk filter, and finally charcoal
Tasting Notes	
Nose	Prominent spice of white pepper, anise, caraway, and fennel; fruit—black cherries, plums, and ripe pears; cocoa
Palate	Robust spice, anise, and licorice giving way to menthol and eucalyptus; layers of roasted nuts, coffee, and chocolate sweetness
Mouth Feel	Rich and full bodied; almost silky; great acidity
Finish	Long and memorable with a touch of bittersweet chocolate
Expert Notes	"Drink it neat, chilled, mixed or just play"; "A well-rounded vodka that would work well in all seasonal cocktails"; "Amazingly complex!"; "Rustic yet elegant"

ERISTOFF

Style and Origin	Old World; Beaucaire, France
Character Score	6.5
Cost	$
Materials & Techniques	Rye; proprietary water source; three-step distillation process; charcoal filtered
Tasting Notes	
Nose	Grainy, bread dough, and yeasty leading to white pepper, vanilla, and cinnamon; hints of lemon and orange peel
Palate	Rich; toasty shades of caramel, toffee, and cocoa; fresh peach and green apple; orange marmalade with a touch of cinnamon and nutmeg spice
Mouth Feel	Rich; silky; medium to full bodied with nice acidity
Finish	Medium; warm yet pleasant; memorable with a touch of bitter chocolate
Expert Notes	"Stirred and served up with just the thinnest slice of orange peel"; "A very civilized glass of vodka"; "Refined and sophisticated… a very well structured spirit"; "Lovely balance, the alcohol supports the flavors in perfect harmony"

PRAVDA

Style and Origin	Old World; Bielsko-Biala, Poland
Character Score	7.5
Cost	$$$
Materials & Techniques	Late harvest "sweet" rye; spring water from the Carpathian Mountains; five-times distilled—patented process; charcoal filtered
Tasting Notes	
Nose	Alcohol forward with quick transition to slightly grainy sweetness; spicy notes of rye bread, caraway, sesame, and white pepper; vegetal notes of asparagus and green bell pepper
Palate	Robust grainy sweetness up front leading to spiciness of black pepper, anise, and ground cinnamon; complimented by dark berries
Mouth Feel	Medium bodied; soft texture with nice acidity
Finish	Medium to long, lingering with the memory of orange zest
Expert Notes	"Feature in a spirit-forward drink"; "Berry Caipiroska"; "Serve straight from the freezer with food—rich, salty, smoked foods"; "Vodka to support a Punch"; "All around a very good vodka"; "Nice structure"

SOBIESKI

Style and Origin	Old World; Starogard Gdanski, Poland
Character Score	8
Cost	$
Materials & Techniques	Dankowski rye; water sourced from Oligocene intakes, softened and demineralized; four-times distilled via continuous distillation; filtration proprietary
Tasting Notes	
Nose	Spicy graininess; fresh-baked rye bread; black pepper and caraway seeds; vegetal earthiness; a touch of lemon peel
Palate	Prominent spiciness; black pepper, cinnamon, clove, cardamom, and allspice giving in to citrus notes of grapefruit, kaffir lime, and lemon
Mouth Feel	Big, fat, rich, and creamy; cools nicely with medium acidity
Finish	Warm; dry; pleasant with lingering memory of crème brûlée
Expert Notes	"This is a vodka that can stand up to vermouth"; "A grapefruit complement—think Greyhound or Ruby"; "Neat"; "Cocktails with Pacific Rim flavors"; "This is real vodka"; "Big yet balanced, very complex"

SQUARE ONE ORGANIC

Style and Origin	Old World; Rigby, Idaho, USA
Character Score	8
Cost	$$$
Materials & Techniques	100 percent Organic Montana rye; well water supplied by the Snake River Plain Aquifer, sourced from 200 feet below the distillery; four-column distillation; single filtration through a natural fiber medium
Tasting Notes	
Nose	Pronounced spice of black pepper, anise, and cinnamon followed by floral notes of honeysuckle and orange blossom giving way to yeasty bread dough, baked apple, and toasted hazelnuts
Palate	Big black pepper followed by vegetal notes of asparagus and green bell pepper leading into a slate-like minerality with sweetness of dried fruits—raisin and fig
Mouth Feel	Slightly oily; viscous; dries nicely with bright acidity
Finish	Warm; medium in length with a pleasant blend of spice and sweet
Expert Notes	"This is good for all cocktails"; "A Bing Cherry Caipiroska"; "Great complement to barbequed smoked sausage"; "A mixologist's vodka!"; "A big, bold mixable vodka that supports fresh ingredients"

WYBOROWA*

Style and Origin	Old World; Poznan, Poland
Character Score	7
Cost	$$
Materials & Techniques	Rye; artesian well water; primary distillation through copper column, then rectification through three stainless steel columns; candle filtration through cellulose
Tasting Notes	
Nose	Earthy, slightly vegetal, floral—lilac, white pepper giving way to grain, bread dough, a hint of vanilla, and citrus of grapefruit and lemon
Palate	Alcohol forward; menthol; lemon oil and anise; traces of white pepper, caramel, toffee, egg cream, and bittersweet cocoa
Mouth Feel	Light to medium bodied with light acidity; slightest oily texture giving in to a dry finish
Finish	Short to medium; staying mostly on the palate with a lingering grainy sweetness
Expert Notes	"Long drinks with fresh-squeezed juice"; "Anything savory, Bloody Mary, Bloody Bull, Caesar"; "I like the character of this vodka, expressive, nicely balanced, a straightforward representation"

*NOT CURRENTLY DISTRIBUTED IN THE UNITED STATES

WYBOROWA EXQUISITE*

Style and Origin	Old World; Poznan, Poland
Character Score	8
Cost	$$$$
Materials & Techniques	Hand-selected, single-estate Dankowskie Zlote rye; spring water; distilled through a single copper column; cellulose plates used for candle-method particle filtration
Tasting Notes	
Nose	Elegant with a pronounced graininess; bread dough; vanilla; fresh-cut grass; citrus of grapefruit; black cherry; green bell pepper and white pepper
Palate	Grainy sweetness; spicy black pepper; lemon rind; candied citrus; bittersweet chocolate; butterscotch
Mouth Feel	Rich, full, creamy, and luscious giving way to a dry finish and very nice acidity
Finish	Long; memorable with a lingering note of toasted Brazil nuts
Expert Notes	"Nice straight over a beautiful cube of ice"; "This is one I want to drink straight, ice cold with my caviar, smoked salmon or duck foie gras"; "Gutsy and robust, not a vodka for the timid!"; "Big, bold and oh so complex, this is what vodka must have tasted like 100 years ago!"

*NOT CURRENTLY DISTRIBUTED IN THE UNITED STATES

WHEAT VODKAS

42 BELOW

Style and Origin	New World, New Zealand
Character Score	6
Cost	$$
Materials & Techniques	100 percent Australian wheat; volcanic spring water; four-times distilled using a "high saturation" technique; charcoal filtered
Tasting Notes	
Nose	Grainy; malted wheat; eucalyptus; mint; spice of anise and coriander; vanilla; lemon peel citrus; red berries with a slight floral note of lily
Palate	Cinnamon, clove and anise spice; vanilla; kumquat and orange; taffy, and sweet butter giving way to minerality and spicy green pepper
Mouth Feel	Viscous; almost oily in texture with light to medium acidity
Finish	Medium in length, exiting with just the faintest touch of spice
Expert Notes	"Great base for Highballs—a Sea Breeze, Greyhound, Cape Cod, or just with tonic water and a lime"; "Would work well in culinary style cocktails featuring flavors of Mexico—jalapeño and cilantro"; "Fairly complex yet easy drinking"

360

Style and Origin	New World; Weston, Missouri, USA
Character Score	7
Cost	$$
Materials & Techniques	Wheat; well water treated by reverse osmosis; four-times distilled via continuous column; five-times filtered, once through granulated coconut shells
Tasting Notes	
Nose	Grain mash with big black pepper; vanilla and anise spice followed by floral hints of lavender and honey blossom
Palate	Alcohol forward then softens, presenting toasted grains, caramel, vanilla, butterscotch, and white chocolate with a touch of eucalyptus
Mouth Feel	Medium bodied; slightly oily; viscous with light to medium acidity
Finish	Short to medium in length; alcohol lingers with white pepper spice
Expert Notes	"A great addition to a Bloody Bull"; "A vodka that stands toe to toe with a good tonic water"; "Robust yet floral and fruity"; "A cocktail vodka for sure"; "Would balance citrus well—lemon, lime, or grapefruit juice"

ABSOLUT

Style and Origin	New World; Åhus, Sweden
Character Score	6
Cost	$$
Materials & Techniques	Swedish winter wheat; privately held well water; continuous distillation; no filtration
Tasting Notes	
Nose	Grainy, cereal, malty, bread dough with white pepper and vanilla giving way to butterscotch, marzipan, and candied orange peel
Palate	Nuttiness of almond and hazelnut; toffee; vanilla; spice of white pepper, anise, and clove; dried fruits—apricot and orange peel
Mouth Feel	Medium weight; rather silky with moderate acidity
Finish	Medium in length; warm and pleasing with just the slightest touch of bitter almond nuttiness
Expert Notes	"Straight up with a twist of orange"; "Nice match for all citrus—perfect for a Harvey Wallbanger or Hurlyburly"; "Great with cream or coffee—White Russian, Mudslide, or Espresso Martini"

ABSOLUT 100

Style and Origin	Old World; Åhus, Sweden
Character Score	8
Cost	$$$
Materials & Techniques	Swedish winter wheat; privately held well water; continuous distillation; no filtration; finished 100 proof
Tasting Notes	
Nose	Alcohol up front; grainy, yeasty bread dough with a stony minerality—wet slate; vegetal leading to lemon rind, vanilla, and floral notes of lavender and lilac
Palate	Vegetal—green bell pepper and asparagus; spicy black pepper, star anise, fennel, cardamom, and hint of eucalyptus followed by stewed dark tree fruits
Mouth Feel	Clean yet full bodied; crisp and dry with very good acidity
Finish	Long; warm yet pleasant with lingering notes of sweet grain and peppery spice
Expert Notes	"Perfect in a Vesper"; "Stirred and straight up with a Maytag blue cheese–stuffed olive"; "Supports savory mixers"; "Great partnered with caviar "; "Very complex, lots going on here, layers of flavor"

ABSOLUT ELYX

Style and Origin	Old World; Åhus, Sweden
Character Score	6.5
Cost	$$$
Materials & Techniques	Hand-selected estate wheat; underground well water; distilled in vintage copper distillation unit; particle filtration
Tasting Notes	
Nose	Sweet cream; allspice; vanilla; pineapple; citrus of lemon and orange; slightly grainy with floral notes of honey blossom
Palate	Malted-toasted grains; sweet creamy butter; toffee and vanilla nougat; a slight slate-like minerality
Mouth Feel	Silky upon entry; dries beautifully leaving a waxy texture with good acidity
Finish	Warm; medium to long; elegant and refined with a touch of toffee sweetness
Expert Notes	"This would be wonderful chilled with a twist of orange, or all by itself straight from the freezer"; "Supports anything with espresso, chocolate, or one big beautiful hand-cut ice cube"; "Balanced, spicy with a light sweetness that goes on and on"

AKVINTA

Style and Origin	Old World; Imotski, Croatia
Character Score	7
Cost	$$$$
Materials & Techniques	100 percent Organic Italian wheat; pure water from the Red and Blue Lakes of Imotski; three-times distilled; five-times filtered—charcoal, marble, silver, gold, and platinum; certified Organic and Kosher
Tasting Notes	
Nose	Earthy notes of mushrooms followed by spice of white and black pepper, allspice, and cinnamon; a touch of thyme with floral notes of lemon verbena
Palate	Spice of pepper, clove and anise; caramel, and vanilla mingled with grapefruit, tangerine, and a slight minerality
Mouth Feel	Comes in round and silky; rather luxurious then dries out and finishes clean with nice acidity
Finish	Medium to long; warm with just a touch of spice
Expert Notes	"A full-figured vodka"; "Serve straight with caviar, oily or smoked fish"; "A mate to ginger—I'm thinking Cucumber Cobbler"; "Nice with fresh lime or orange juice"; "Robust, complex yet approachable"

CHOPIN—WHEAT

Style and Origin	Old World; Krzesk-Majatek, Poland
Character Score	6
Cost	$$$
Materials & Techniques	Winter wheat from the Podlasie region of Poland; purified well water; four-times distilled—initially in a copper column, then rectified three times in rectification columns; five-times filtered—plate filter, two candle filters, disk filter, then charcoal
Tasting Notes	
Nose	Opens up with bread-dough graininess leading to white pepper and fennel spice; herbal and vegetal expressions follow with just the memory of lemony citrus and orange blossom
Palate	Spicy with pepper—more black than white—and ground cloves; rich butterscotch, caramel, vanilla, and toasted walnuts leading to dark tree fruits and a touch of lime
Mouth Feel	Medium weight; clean; well balanced with lovely acidity
Finish	Medium to long; clean finish with lingering hints of toasted nuts
Expert Notes	"Does not need anything but a glass and a tin of Ossetra caviar...straight from the freezer"; "There is a whole lot going on in this glass, it continues to express itself"; "Great with fresh juices"; "Good, clean distillate with nice complexity"

DEATH'S DOOR

Style and Origin	Old World; Madison, Wisconsin, USA
Character Score	7.5
Cost	$$$
Materials & Techniques	Two varieties of red, hard winter wheat; water source proprietary; triple distilled in a Christian Carl hybrid pot and column still; filtration proprietary
Tasting Notes	
Nose	A big, complex nose that opens with an herbal, yeasty, grainy appeal followed quickly by big spice—anise, fennel, black pepper, and coriander—with a hint of briny minerality
Palate	Black tea complimented by a blast of spice—cardamom, clove, and black pepper; yeasty baked bread; roasted fennel and baked apple
Mouth Feel	Robust and slightly oily with nice acidity
Finish	Medium in length leaving a pleasant, slightly spicy reminder
Expert Notes	"This is a Bloody Bull vodka for sure"; "Neat with smoked fish, salmon, caviar, or ceviche"; "Pair with anything savory, this is a food vodka!"; "Big character with good balance—savory, spicy, complex all in harmony"

DOUBLE CROSS

Style and Origin	Old World; Slovak Republic
Character Score	7
Cost	$$$$
Materials & Techniques	Winter wheat; Tatra mountain spring water; seven-times column distilled; seven-times filtered through active charcoal, limestone, and other materials
Tasting Notes	
Nose	Citrus forward—sweet orange and lemon zest; black and white pepper; toasted grain with cream soda, vanilla, cut straw, and green apple
Palate	Prominent white pepper followed by sweet vanilla, caramel, butter cream, clove, ginger, dark fruits, berries, and candied orange peel
Mouth Feel	Rich in the mouth; silky; almost oily texture with very nice acidity
Finish	Medium to long with a clean lingering memory of white pepper and orange rind
Expert Notes	"Great over ice with a twist of lemon or orange"; "Match with food—caviar, the raw bar, or smoked fish"; "Excellent with fresh berries in a Caipiroska"; "A joy to drink!"

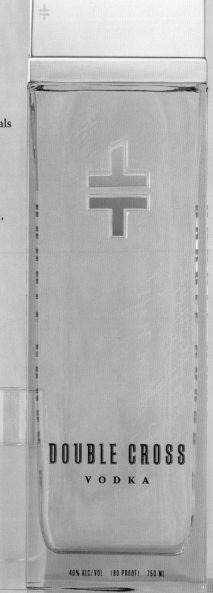

EFFEN

Style and Origin	New World; Hoojhoudt, Netherlands
Character Score	6.5
Cost	$$$
Materials & Techniques	Wheat from Northern Holland; water sourced from deep natural aquifers of Northern Holland, treated with ultraviolet light then filtered through a high-pressured reverse osmosis system; distilled in a four-column still combined with a proprietary process allowing distillation at lower temperatures; twice filtered through activated carbon
Tasting Notes	
Nose	Citrus forward primarily orange; bread dough; vanilla; toasted hazelnuts; white pepper; pear and apple; floral highlights of orange blossom; hint of green bell pepper
Palate	Enters with honey-butterscotch sweetness, vanilla and blanched almonds giving way to prominent citrus of grapefruit and lemon
Mouth Feel	Medium bodied; slightly waxy with mild acidity
Finish	Short to medium; lingers with a bittersweet nuttiness
Expert Notes	"Lends itself to fresh, fruit-forward cocktails"; "This vodka would play nice with sushi"; "Nice structure, balance, and complexity"; "Has a natural sweetness that would pair nicely with chocolate or coffee"

FRÏS

Style and Origin	Old World; Aalborg, Denmark
Character Score	7
Cost	$
Materials & Techniques	Wheat; a blend of well waters from both Aalborg, Denmark, and Fort Smith, Arkansas, USA—both sources purified through reverse osmosis; continuous six-column distillation combined with freeze distillation; no filtration
Tasting Notes	
Nose	Bread dough; buttery; caramel; malted milk; vanilla forward; cinnamon; orange blossom, and lavender rounding out citrus notes of sweet orange
Palate	Prominent vanilla; rich egg cream; white pepper spice; baked apple; tropical fruits
Mouth Feel	Rich; soft; coats the palate initially then exits clean with medium acidity
Finish	Medium to long finish that starts rather warm but cools off nicely with just a touch of menthol and white pepper
Expert Notes	"Would make a great Vodka Gimlet"; "Great for long drinks—think refreshing summertime coolers or Pink Sangria"; "Crowd pleaser"; "Very nicely balanced, nice structure with pronounced flavors"

GREY GOOSE

Style and Origin	New World; Picardy, France
Character Score	5.5
Cost	$$$
Materials & Techniques	French, soft winter wheat; spring water from the town of Gensac in the Champagne region of Cognac, France; five-step distillation process; active carbon and cellulose filtration
Tasting Notes	
Nose	Grainy; yeasty; bright citrus of both orange and orange peel; spicy ground white pepper; fresh green bell pepper with floral notes of lavender and fresh-cut grass
Palate	Toffee sweetness; touch of vanilla; sweet butter; cream soda; blanched almond; licorice and white pepper
Mouth Feel	Rich; slightly viscous; round texture; coats the palate with nice acidity
Finish	Medium to long finish with just the memory of anise and almond
Expert Notes	"Seasonal, fruit-forward cocktails"; "Vodka Gibson"; "Simply neat over ice"; "Good for Collinses or Sours, a good base for citrus driven drinks"; "Approachable with nice character, nicely balanced"

IMPERIA

Style and Origin	Old World; St. Petersburg, Russia
Character Score	7
Cost	$$$
Materials & Techniques	Winter wheat; soft glacial water from Lake Ladoga; eight-times distilled; four-times birch-charcoal filtered followed by crystal quartz filtration
Tasting Notes	
Nose	Yeasty; cooked grains; porridge with bananas and cinnamon; slightly malty; hints of vanilla bean; sweet orange
Palate	Starts with a grainy sweetness; minerality; spice—anise, white and black pepper—then moves into cinnamon, toffee, and chocolate with a hint of lime peel
Mouth Feel	Silky; slightly oily, which coats the palate; nice acidity
Finish	Medium in length with a nice warm glow and the faintest memory of menthol
Expert Notes	"Vodka for all occasions, from sipping neat to cocktails"; "Muddle nicely with citrus and fresh berries"; "Vodka cocktails want to be made with this"; "Beautifully balanced—stands on it's own"

JEAN-MARC XO

Style and Origin	New World; Cognac region of France
Character Score	6
Cost	$$$$$
Materials & Techniques	Four varieties of French wheat: Ysengrain, Orvantis, Azteque, and Chargeur; pure Gensac spring water, naturally filtered through Grande Champagne limestone; nine-times distilled in French Alembic copper pot stills; micro-oxygenated; Limousin oak charcoal filtered
Tasting Notes	
Nose	Very floral—jasmine, hibiscus, and lemon verbena, with eucalyptus layered upon notes of cinnamon, vanilla, and clove with a yeasty bread-dough underbelly
Palate	French toast with cinnamon; vanilla and caramel with toasty nuttiness; slightly vegetal—hints of rue
Mouth Feel	Elegant, luscious, silky texture with light to medium acidity
Finish	Long, refined, and delicate with a lingering touch of baking spice
Expert Notes	"This is vodka for culinary-inspired cocktails"; "Works well with citrus"; "Would make a nice after-dinner vodka—a digestif"; "A chef's vodka"; "Nicely balanced, complex yet approachable—will be great all on its own!"

KETEL ONE

Style and Origin	New World; Schiedam, Holland, Netherlands
Character Score	6
Cost	$$
Materials & Techniques	Winter wheat from Holland and Northern France; reverse osmosis purified water; distillation using a combination of both pot and column stills; gravity-driven charcoal filtration
Tasting Notes	
Nose	Alcohol supports a vibrant, grainy, and slightly yeasty bread-dough character; vanilla; cinnamon; green apple; hint of sweet orange
Palate	Rich with toasty grains, highlighted by touches of vanilla, marzipan, baked apple pie, and cream soda
Mouth Feel	Medium bodied; lightly viscous with very good acidity
Finish	Medium in length; warm with a lingering grainy sweetness
Expert Notes	"Drink neat from the freezer paired with gravlax"; "Good for mixology cocktails as well as a great foundation for classic vodka drinks"; "Straightforward, balanced, and approachable"; "A grain-driven vodka"

PEARL

Style and Origin	New World; Alberta, Canada
Character Score	7
Cost	$
Materials & Techniques	Winter wheat; pure Canadian Rocky Mountain spring water; five-times distilled; six-times filtered
Tasting Notes	
Nose	Bread dough; malted grains leads into vanilla, white pepper, and citrus—orange blossom, orange, and lemon oil; hint of asparagus and minerals reminiscent of wet river stones
Palate	Cream soda; caramel; honey; blanched-almond nuttiness; baking spices of clove and allspice with sweet orange
Mouth Feel	Soft; slightly creamy with nice viscosity and pleasing acidity
Finish	Warm; medium in length with a pleasant memory of white pepper and sweet vanilla
Expert Notes	"Play with this one—maybe fresh herbs or tropical fruits"; "An awesome Moscow Mule"; "Try this one in an Apple Blossom"; "Dessert cocktails"; "Big flavor, lots to work with—robust yet balanced"

PURUS

Style and Origin	New World; Lagnasco, Italy
Character Score	6
Cost	$$$
Materials & Techniques	100 percent Organic Italian wheat; water sourced from the Italian Alps; five-times distilled; active-charcoal filtration
Tasting Notes	
Nose	Big cereal graininess and bread dough, coupled with shades of fresh citrus—lemon and orange; honeydew melon; a slight minerality—wet stone; touches of vanilla, licorice, cinnamon, and white pepper
Palate	Vanilla, white chocolate, and egg cream with tropical fruits, ripe melon, and a touch of spice from ground white pepper and anise
Mouth Feel	Buttery, rich, and slightly oily with medium acidity
Finish	Medium in length with a warm and memorable lasting impression
Expert Notes	"Neat from the freezer, over ice, or in a Martini"; "Nice in a Fizz"; "Beautiful distillate"; "A great example of an approachable, yet complex, clean, crisp spirit"; "Intriguing"

RUSSIAN STANDARD

Style and Origin	Old World; St. Petersburg, Russia
Character Score	7.5
Cost	$$
Materials & Techniques	Winter wheat from the Russian Steppes; water sourced from Lake Ladoga; four-times column distilled—with the entire process repeated a second time; charcoal filtration
Tasting Notes	
Nose	Sweet graininess; bitter orange giving way to floral notes of honeysuckle; slight limestone minerality; vanilla and cracked black pepper
Palate	Malted graininess; vanilla; caramel; almond; anise; black pepper; a touch of lemon rind
Mouth Feel	Rich and creamy; round; almost oily with nice acidity
Finish	Medium in length; warm; vibrant with rich notes of marzipan
Expert Notes	"Great base for cocktails"; "Classic, traditional, full of flavor and character—you know you are drinking vodka"; "Elegant alongside foie gras and a sprinkle of sea salt"

SHPILKA

Style and Origin	Old World; Kyrgyzstan
Character Score	7
Cost	$$
Materials & Techniques	Russian winter wheat; Artesian spring water; six-times distilled; filtered through five natural elements, including charcoal, quartz, and pumice
Tasting Notes	
Nose	A slightly grainy nose; yeast and bread dough giving way to citrus—both sweet and bitter orange—with lemon zest; green apple; white pepper; cucumber skin and a flint-like minerality
Palate	Layers of cereal graininess; sweet toffee; nougat; vanilla and black pepper spice; herbaceous subtleties; a touch of minerality and dried stone fruit
Mouth Feel	Rich velvety texture that dries nicely with good acidity
Finish	Medium in length; comes in slightly warm then cools nicely with a touch of sweetness
Expert Notes	"Rocks with a peel of lemon, but would support even the subtlest of ingredients"; "It would be lovely straight out of the freezer!"; "This vodka wants to be alone"; "A clean straightforward vodka"

SHPILKA
ШПИЛЬКА

VODKA
ВОДКА

IMPORTED
DISTILLED FROM RUSSIAN WHEAT
750 ML   40% ALC./VOL.

SVEDKA

Style and Origin	Old World; Lidköping, Sweden
Character Score	8
Cost	$
Materials & Techniques	Swedish winter wheat; water sourced from the "Råda ås" hills and Lake Vanern; five-times distilled; light carbon filtration
Tasting Notes	
Nose	Toasted grains; lime peel and green apple; herbaceous elements—cilantro and thyme; vegetal notes of green bell pepper; vanilla; wet limestone
Palate	Pronounced alcohol supporting butter-scotch, caramel and crème brûlée; spice of cinnamon, allspice, white pepper and star anise; baked apple and pear
Mouth Feel	Rich; medium bodied; satin almost silky; warm but cools slowly with pronounced acidity
Finish	Medium in length with a lingering touch of bittersweetness
Expert Notes	"Support big flavors—Bloody Bull, Caesar"; "You know you are drinking vodka!"; "Rustic and powerful!"; "Robust—will stand up to rich savory flavors"; "If it's the Holidays, I'm thinking Glögg"

TANQUERAY STERLING

Style and Origin	New World; Scotland, UK
Character Score	6.5
Cost	$$
Materials & Techniques	Wheat; demineralized water; three-times pot distilled; filtration through silver impregnated carbon
Tasting Notes	
Nose	Grainy; malty; yeasty with an earthy quality and big baking spices—anise, allspice, ginger, clove, and white and black pepper spice; floral notes of lemon verbena
Palate	Butterscotch; vanilla and mocha; black pepper; ginger; stewed dark fruits—cherry and plum; lemon peel and minerality
Mouth Feel	Enters with rich viscosity then dries beautifully with bright acidity
Finish	Medium to long; memorable with a lemony tingle and butterscotch sweetness
Expert Notes	"A beautiful platform to build flavors upon"; "Lemon Spot, Cosmopolitan, and Kamikaze"; "A robust spirit that shows its alcohol proudly without overwhelming the drinker!"; "Finishes with a lasting impression"; "Well balanced"

VAN GOGH BLUE

Style and Origin	New World; Schiedam, Holland
Character Score	5.5
Cost	$$$
Materials & Techniques	Blend of wheat from France, Germany, and Netherlands; water treated with reverse osmosis; triple distilled using combination of column and pot; cartridge filtration
Tasting Notes	
Nose	Grains up front—bread dough and porridge; fresh citrus—lemon, lime, and orange; vanilla; white pepper spice; herbaceous notes of dried oregano
Palate	Caramel, toffee, toasted nuts, and pronounced vanilla leading to green bell pepper and asparagus, creamed corn, and a hint of black pepper
Mouth Feel	Rich; almost lanolin in texture with very good acidity
Finish	Medium in length; warm with a lingering spiciness of white pepper and blanched almonds
Expert Notes	"A vodka well suited for tonic. Garnish with a slice of cucumber—yum"; "Correct, straightforward vodka with great versatility"; "Beautiful balance with just a touch of sweetness"

POTATO VODKAS

BLUE ICE

Style and Origin	New World; Rigby, Idaho, USA
Character Score	6
Cost	$$
Materials & Techniques	Idaho Russet Burbank potatoes; water source—naturally filtered through volcanic rock from Rocky Mountain glacial and snow melt; four-column fractional distillation; five-step filtration via charcoal, filter press, garnet or crystal, travertine, and submicron
Tasting Notes	
Nose	Earthy; herbaceous; baked potato; lemon peel and oil; white pepper and star anise
Palate	Black pepper; sweet vanilla, caramel and toffee; ginger; citrus of orange, lemon and kumquat; a hint of minerality
Mouth Feel	Richly textured; medium bodied and slightly creamy with nice acidity
Finish	Medium; warm and lingering with a nice spicy kick
Expert Notes	"This is a food vodka—caviar or smoked fish"; "Drink this straight up with a blue cheese–stuffed Spanish olive"; "Would be nice chilled and straight"; "Elegant and refined with beautiful structure"; "Everything is in harmony"

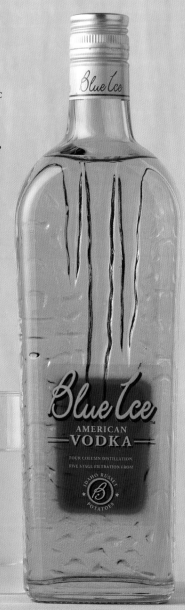

BOYD & BLAIR

Style and Origin	Old World; Glenshaw, Pennsylvania, USA
Character Score	7.5
Cost	$$$
Materials & Techniques	Potato; water treated with reverse osmosis and UV light; triple distilled with a combination of pot and column methods; standard carbon filtration followed by fixed-bed filtration through coconut-shell carbon
Tasting Notes	
Nose	Big and complex with an intriguing up-front aroma of peeled, uncooked potatoes followed by rich custard, rice pudding, caramel, toffee, and ripe tropical fruits
Palate	Layers of flavors beginning with banana-nut bread, toffee, sweet vanilla, caramel, and cocoa followed by ripe tropical fruits— mango, papaya, and coconut
Mouth Feel	Luscious, silky, and elegant with light acidity
Finish	Medium to long; lingering with the memory of bittersweet chocolate and honey
Expert Notes	"Serve it straight with Pacific Rim dishes"; "This would make a tasty Chi-Chi!"; "Would balance beautifully with fresh citrus"; "Creamy, elegant, rich!"; "It's lovely, complex, fascinating, and very unique!"

CHOPIN—POTATO

Style and Origin	Old World; Krzesk-Majatek, Poland
Character Score	7
Cost	$$$
Materials & Techniques	Stobrawa potato; purified artesian well water; four-times distilled—copper column, then three times in rectification columns; filtered five times—initial plate, two candle filters, disk, then charcoal
Tasting Notes	
Nose	Mashed potato; mushroom; toasted nuts; anise; orange peel with a vegetal grassy quality highlighted by vanilla bean and white pepper
Palate	Assertive character; a touch of sweetness up front followed by white pepper spice; vegetal hints of green bell pepper; slight minerality
Mouth Feel	Soft, rich, and velvety with notable acidity
Finish	A lasting and long finish, ending with a touch of bitter almond and slate
Expert Notes	"This is an all-purpose vodka that will complement a full range of drinks"; "Straight out of the freezer"; "Bring on the caviar"; "A lovely glass of vodka, a wonderful representation of this style"; "Crowds will love it!"; "Powerful yet beautifully balanced"

CHRISTIANIA

Style and Origin	Old World; Norway
Character Score	7.5
Cost	$$$$
Materials & Techniques	Trondeleg Organic potatoes; arctic spring water; six-times distilled; charcoal filtration then aerated
Tasting Notes	
Nose	Bright citrus of grapefruit and lemon leading to green apple and melon; herbal notes of thyme and lemon verbena; light white pepper spice, and earthy mashed potato
Palate	Musty and earthy featuring potato and vegetal notes of green bell pepper; white pepper spice; sweet cocoa, and just a touch of minerality
Mouth Feel	Silky; round and creamy; cools nicely, leaving a pleasant waxy sensation on the palate with medium acidity
Finish	Medium with just the lingering memory of cocoa
Expert Notes	"Martini vodka"; "This is a hearty vodka that wants to be dirty!"; "This vodka stands up—straight from the freezer"; "Perfect vodka to support culinary-style cocktails"; "A refined, complex, elegant vodka"

KARLSSON'S GOLD

Style and Origin	Old World; Cape Bjäre, Sweden
Character Score	7
Cost	$$$
Materials & Techniques	Seven varieties of new potatoes, harvested prior to the skin's formation; Swedish water made neutral through reverse osmosis; single continuous distillation; particle filtration only
Tasting Notes	
Nose	Caramel; vanilla; cream soda; cocoa; bittersweet chocolate; a touch of pecan nuttiness with vegetal, earthy tones
Palate	Dark chocolate; caramel corn; baked apple with brown sugar and cinnamon; malted milk
Mouth Feel	Full; rich; coats the palate with moderate acidity
Finish	Medium to long; pleasant with a memorable nod of dark chocolate
Expert Notes	"It seems to want food—cheese perhaps?"; "A vodka for the discerning vodka drinker"; "An interesting nose that follows with complexity on the palate"; "After-dinner cocktail or featured in desert-style cocktails"; "A big, full-bodied spirit"

LUKSUSOWA

Style and Origin	Old World; Poznan, Poland
Character Score	7
Cost	$$
Materials & Techniques	Potatoes from Poland's northern Baltic coast; artesian well water sourced at the distillery; triple distilled—first, copper then stainless steel columns; filtration through cellulose-plated activated carbon then candle filtered
Tasting Notes	
Nose	Slightly vegetal—asparagus, baked potato skins; earthy with hints of fennel, anise, and white pepper
Palate	Big pepper spice with traces of sweet vanilla and cocoa, lemon peel, and blanched almonds
Mouth Feel	Rich; medium bodied; silky with bright acidity
Finish	A warm, medium finish with lingering traces of spice
Expert Notes	"An all-purpose vodka"; "Drink it chilled, served with food"; "Consider this paired with caviar"; "Great straight from the freezer with some pickled herring"; "Shows like a classic Eastern European vodka"

MIXED-GRAINS VODKAS

DRAGON BLEU

Style and Origin	New World; Saint-Preuil, France
Character Score	6.5
Cost	$$$
Materials & Techniques	Blend of wheat, barley, and rye; Gensac spring water; small batch micro-distilled; filtration proprietary
Tasting Notes	
Nose	Spicy green pepper; asparagus; soft citrus with toasty grains; an underlying toffee and maple sweetness; a floral note of orange blossom
Palate	The palate mimics the nose along with a soft sweet fruitiness, a touch of white chocolate, egg cream, vanilla, and toasted hazelnuts
Mouth Feel	Clean yet slightly silky with nice acidity
Finish	Medium in length; clean; dry with the slightest note of bitter almond
Expert Notes	"Summer fruit cocktails—watermelon would be tasty"; "A great vodka for cocktails that require a solid foundation"; "Unique"; "This vodka grew on me; I liked it more with each sip"

THE JEWEL OF RUSSIA CLASSIC

Style and Origin	Old World; Russian Federation
Character Score	7.5
Cost	$$$
Materials & Techniques	Russian hard winter wheat and rye; deep artesian well water; four-times distilled; five-step filtration by gravity through quartz sand, birch wood charcoal, and cellulose
Tasting Notes	
Nose	Cereal; toasted grains; spices of cinnamon, clove, and white pepper; sweet vanilla; green apple; licorice; slight citrus notes of lemon and orange
Palate	Powerful entry with lots of spice of anise, clove and black pepper; lemon and orange peel; caramel; toffee and Mexican chocolate
Mouth Feel	Rather viscous; slightly oily with pleasing acidity
Finish	Very long; lingering and warm with just a faint memory of toasted almond
Expert Notes	"Would support anything savory"; "Serve neat, ice cold with cured meats"; "Would be great served simply over ice or with soda water and a twist"; "A nice vodka for infusions"; "Versatile and classic"

PURITY

Style and Origin	Old World; Ellinge Castel, Sweden
Character Score	6.5
Cost	$$$$
Materials & Techniques	Mix of certified Organic winter wheat and barley; both natural untouched water and deionized water from the same proprietary source; copper pot still combined with twin column distillation; no filtration
Tasting Notes	
Nose	Bread dough; malty; sweet vanilla; caramel; citrus of lemon and lime; honeydew melon; mango and pineapple; floral notes of orange blossom and honeysuckle
Palate	Malty; grainy sweetness; tree fruits of cherry and plum; stewed red fruits; lime peel; raisins with white pepper and vanilla
Mouth Feel	Round, soft, creamy, and slightly oily with lovely acidity
Finish	Medium in length; clean with a sweet nougat reminder
Expert Notes	"Great mixed with red fruits —cherries, red raspberries, cranberries"; "H2O cocktails"; "Will work well in Bond's Vesper"; "A real crowd pleaser"; "A stand-up-and-be-noticed vodka"

REYKA

Style and Origin	Old World; Borgarnes, Iceland
Character Score	7.5
Cost	$$
Materials & Techniques	Wheat and barley; pure Icelandic spring water from Grabok Spring; distillation via unique Carter Head still; lava rock filtration
Tasting Notes	
Nose	Pronounced alcohol with bold spiciness of anise, cinnamon, and black pepper; hint of vanilla with underlying graininess
Palate	Enters with cereal graininess; layers of toffee, caramel, and cream soda give way to spicy white pepper and star anise; slight earthy mushroom element
Mouth Feel	Full bodied; slightly creamy with medium acidity
Finish	Medium in length; lingers with slight aniseed and white pepper spiciness
Expert Notes	"Bloody Bull"; "This is a powerful vodka that will stand up to a big Spanish olive"; "Good balance for the Holiday spices in Glögg"; "Comes in assertively, yielding a very bold character"

STOLICHNAYA

Style and Origin	Old World; Riga, Latvia
Character Score	7
Cost	$$
Materials & Techniques	Wheat and rye; water from the Kaliningrad region; three-times distilled; quadruple filtration
Tasting Notes	
Nose	Sweet orange up front giving way to toasty grain, caramel, cocoa, and vanilla with subtle hints of apple and ripe pear
Palate	Alcohol forward yet balanced; a sweet underlying graininess supporting flavors of sweet vanilla, butterscotch, and egg cream with orange rind
Mouth Feel	Rich, honeyed, and full bodied with mild acidity
Finish	Short to medium in length, leaving traces of citrus and bread pudding
Expert Notes	"This would own a Vodka Tonic"; "In a Cape Cod or Cosmopolitan"; "Mixed with any fruit juice—orange in particular"; "Use for your Harvey Wallbanger"; "Drink it neat, well chilled"

STOLICHNAYA ELIT

Style and Origin	Old World; Riga, Latvia
Character Score	6.5
Cost	$$$$$
Materials & Techniques	Wheat and rye; artesian well water; triple distilled; freezing (0° Fahrenheit) filtration system
Tasting Notes	
Nose	Alcohol present, with touches of green bell pepper, cereal grain, and fresh citrus—orange, lemon, and tangerine; hints of vanilla, white pepper, anise, and mint
Palate	Presents with bread dough layered with creamy caramel and butterscotch; toffee coupled with sweet vanilla and bright citrus
Mouth Feel	Lovely mouth feel; great texture; clean; crisp; rather luxurious; medium acidity
Finish	Medium in length; memorable, leaving just a memory of anise spice and caramel sweetness
Expert Notes	"All it needs is a glass—perfect for a Vodka Martini"; "Elegant and complex; drink it on its own"; "Refined vodka"; "With a simple twist of lemon and nothing more"

ULTIMAT

Style and Origin	Old World; Bielsko-Biala, Poland
Character Score	6.5
Cost	$$$$
Materials & Techniques	Wheat, rye, and potato; purified, reverse osmosis filtered water; column distilled, rectified through seven steel columns; copper filtration—four times, then a fifth through activated carbon
Tasting Notes	
Nose	Rather floral and herbaceous with nice minerality; notes of lemon peel and spice—white pepper, clove, and cinnamon; supported by a grainy, buttery, almost pastry-dough underbelly
Palate	Vanilla; cream soda; ground clove; pear and apricot; caramel; butterscotch; cocoa, and rice pudding
Mouth Feel	Medium bodied; silky in texture; slightly oily; very nice acidity
Finish	Medium in length; warm but cools nicely with a touch of pepper and bitter almond
Expert Notes	"Round, fun, drinkable; it would work with anything"; "Good all-purpose vodka"; "Perfect for a Black Boot or Cocoa à Trois"; "Complex, refined, sophisticated"

ZYR

Style and Origin	Old World; Rayazon, Russia
Character Score	7
Cost	$$$
Materials & Techniques	Winter wheat and rye; local well water filtered five times; small batch production, five-times distilled; four-stage filtration, including sand and birch charcoal
Tasting Notes	
Nose	Spicy white pepper; sweet orange; clove; coriander and anise; cereal graininess; dried tarragon; floral notes of honey blossom
Palate	Spicy white pepper followed by sweet orange; spice of clove, coriander and anise; cereal graininess; dried tarragon; floral notes of honey blossom
Mouth Feel	Rich; medium bodied; nicely viscous with good acidity
Finish	Warm; medium in length; lingering and pleasantly clean
Expert Notes	"Straight up with a twist of orange"; "Think Vesper, Vodka Gibson, or the Wizard"; "Very versatile"; "Elegant while paying tribute to its Old-World character"

CORN VODKAS

CROP ORGANIC

Style and Origin	New World; West Central, Minnesota, USA
Character Score	5.5
Cost	$$
Materials & Techniques	Organic corn; Minnesota water—carbon treated then purified through a reverse osmosis system; distillation process is proprietary; no filtration; certified Organic
Tasting Notes	
Nose	Grainy; light toast; cut hay; lemon peel; slightly earthy; green bell pepper; hint of minerality
Palate	Opens with a nice grainy sweetness; creamed corn; ripe cantaloupe and baked apple with citrus of kumquat and orange
Mouth Feel	Medium bodied texture; slightly silky; dries nicely with light acidity
Finish	Warm; medium in length with a sweet and grainy ending
Expert Notes	"Caipiroska"; "Would mix well with tart lemonade"; "Vodka Collins"; "This would be my Moscow Mule vodka"; "A vodka that wants to be mixed"; "Smooth and understated"

MOON MOUNTAIN

Style and Origin	New World; Lawrenceburg, Indiana, USA
Character Score	4
Cost	$$
Materials & Techniques	Organic Midwestern corn; purified water; copper pot distillation—triple distilled; filtration proprietary
Tasting Notes	
Nose	Delicate yet complex with floral notes of honey blossom; cooked grain mash; vanilla, caramel, butterscotch, and egg cream with a note of white pepper
Palate	Maple syrup sweetness followed by cracked black pepper spice; essence of orange oil and a touch of bittersweet chocolate
Mouth Feel	Light bodied; slightly oily upon entry then finishes dry with nice acidity
Finish	A warm entry that cools nicely; finishes rather quickly leaving the memory of bittersweet chocolate
Expert Notes	"Simply with soda and a slice of orange"; "This vodka wants acid; Sours, Collins, or a Cosmopolitan"; "Would support espresso, chocolate, vanilla, caramel—Cocoa à Trois"; "Very mixable"

PRAIRIE ORGANIC

Style and Origin	New World; Benson, Minnesota, USA
Character Score	6
Cost	$$
Materials & Techniques	Organic yellow corn; water source proprietary; distilled a minimum of four times; filtration proprietary; certified Organic and Kosher
Tasting Notes	
Nose	Fresh-cut grass; grainy; creamed corn; ripe tropical fruits and sweet melon; citrus of sweet orange and bright lemon peel
Palate	Alcohol is subtle but present, revealing layers of sweet vanilla, cream soda, ripe mango, and orange alongside a spicy kick of white pepper and ground cinnamon
Mouth Feel	Rich, luscious, and silky with pleasing acidity
Finish	Medium to long; lingers nicely with the memory of warm buttered cornbread
Expert Notes	"Will go great with strawberry or lychee"; "Mixes well with any and all fresh, seasonal ingredients"; "Iced tea, lemonade, or straight from the freezer"; "Easy to drink"; "Well done"

RAIN ORGANICS

Style and Origin	New World; Frankfort, Kentucky, USA
Character Score	6
Cost	$$
Materials & Techniques	Fizzle Flats Farm Organic white corn; Kentucky limestone water; seven-step distillation process; filtered as needed; certified Organic
Tasting Notes	
Nose	Big, sweet vanilla and caramel verging on caramel corn; cream soda married to egg cream with hints of dark chocolate and ground cinnamon
Palate	Everything on the nose comes through on the palate—vanilla, caramel, butterscotch, chocolate, and rich custard—with a slight nod to citrus of sweet orange and ripe mango
Mouth Feel	Medium bodied; silky and creamy but finishes rather dry with pleasing acidity
Finish	Medium to long; warm with sweet caramel lingering on and on
Expert Notes	"This would make a delicious Espresso Martini"; "Excellent infusion backdrop"; "Take it one step further and consider this as balance for savory—Caesar or Bloody Mary"; "Soft, balanced, unique, with a hint of sweetness that dances on the palate"

SMIRNOFF NO. 21

Style and Origin	New World; United States
Character Score	6
Cost	$
Materials & Techniques	Corn; water source proprietary; triple distilled; ten-times filtered through activated charcoal
Tasting Notes	
Nose	A slight tickle of alcohol with vegetal notes of green bell pepper followed by sweet creamed corn, vanilla, caramel, and toasty grains
Palate	Vanilla, caramel, cocoa, and butterscotch sweetness up front with just a touch of white pepper spice followed by sweet orange
Mouth Feel	Soft, rich, and slightly creamy with balanced alcohol and moderate acidity
Finish	Medium in length; warm; clean notes of grainy sweetness
Expert Notes	"Very mixable"; "Fresh fruit cocktails"; "Pleasant, consumer friendly"; "Moscow Mule or Straight up with a twist of lemon"; "would make a fine Vodka & Tonic"; "Nice match for the Ruby"

TITO'S

Style and Origin	New World; Austin, Texas, USA
Character Score	6.5
Cost	$$
Materials & Techniques	Corn; limestone water; six-times pot distilled; carbon filtration
Tasting Notes	
Nose	Herbaceous and vegetal with notes of radish; nice minerality; spice of cinnamon, white pepper, anise, and fennel with lemon-peel citrus
Palate	Pepper spice followed by red tree fruit of black cherry and plum; tropical notes of grilled pineapple; creamed corn, and a touch of toffee sweetness
Mouth Feel	Full, rich, round and creamy; finishes dry with good acidity
Finish	Medium to long; leaves a lasting warmth and just the slightest note of nuttiness
Expert Notes	"Would make a great Salty Dog or Vodka Tonic"; "Stirred and served straight with a couple of pearl onions"; "Mix it with pomegranate juice"; "Make a mint lemonade"; "This one keeps revealing itself with layers of complexity"

OTHER VODKAS

BROKEN SHED

Style and Origin	New World; Central Otago, New Zealand
Character Score	3.5
Cost	$$$$
Materials & Techniques	Whey; water sourced from the Southern Alps region of New Zealand; four-stage distillation; three-stage filtration; batch production
Tasting Notes	
Nose	Cookie dough; both orange and lemon zest; green apple; honeydew melon; white pepper, vanilla, and ground fennel seeds
Palate	Caramel, sweet cream, and butterscotch; citrus notes of blood orange; a touch of mint and black pepper
Mouth Feel	Lightly viscous; dissipates quickly, leaving a clean finish with light to medium acidity
Finish	Medium in length; warm and slightly spicy
Expert Notes	"Highballs"; "This is a vodka to use with fresh herbs in cocktails"; "Straightforward, very mixable"; "A good match for cream-based drinks—the Mudslide, for sure"

CÎROC

Style and Origin	New World; France
Character Score	4
Cost	$$$
Materials & Techniques	Mauzac and Ugni Blanc grape varieties from the Gaillac region of France; water source proprietary; distilled five times —the fifth in a copper pot still; filtration proprietary
Tasting Notes	
Nose	A huge blast of citrus initially—orange, lemon, mandarin, and lime—followed by a fresh grape mash; slight minerality, and a lilac floral waft; alcohol remains present, supporting all aromas
Palate	Grape-forward fruity character followed by a touch of citrus pith and lemon oil
Mouth Feel	Light to medium; slightly silky with moderate acidity
Finish	Short to medium; clean yet with the lasting memory of citrus—primarily lemon zest
Expert Notes	"Lemon Drop or Cosmopolitan"; "Would support anything with fresh berries"; "This is a good vodka to reach for when you need strong citrus character"; "A unique vodka with its own distinctive characteristics"

FAIR

Style and Origin	New World; Cognac, France
Character Score	6
Cost	$$$
Materials & Techniques	Quinoa sourced from fair trade certified co-op farmers of the Bolivian highlands; private spring water filtered through chalk from the Cretaceous age; column distillation; cellulous filtration
Tasting Notes	
Nose	Fruit forward—baked apple, green apple, and stewed fruit; slightly vegetal; mushroom earthiness; finely ground black pepper; finishes with caramel, cocoa, and butterscotch
Palate	Grainy cereal sweetness; stewed blackberries and raspberries; spicy green pepper giving way to minty menthol; hint of strong brewed black tea
Mouth Feel	Light to medium bodied; dries nicely with light acidity
Finish	Short; warm with a bitter almond nuttiness on the back palate
Expert Notes	"This would be my Screwdriver, Greyhound, or Madras vodka"; "Has a unique, complex, and layered nose that continues to express itself"; "Very pleasant, big flavor, unique"

FINLANDIA

Style and Origin	New World; Koskenkorva, Finland
Character Score	5
Cost	$
Materials & Techniques	Finnish six-row barley; glacial spring water from Rajamaki, Finland—filtered through a glacial moraine; seven-step, continuous multipressure column distillation; filtration proprietary
Tasting Notes	
Nose	Spice first with white pepper, caraway, and coriander; grassy; grainy; bright citrus of sweet orange and lemon peel; floral note of chamomile
Palate	Ripe honeydew melon; green apple; orange pith; hint of caramel sweetness, white pepper, vanilla, and allspice
Mouth Feel	Medium body; slightly silky with very good acidity
Finish	Medium long; warm yet friendly with a touch of grain
Expert Notes	"With tonic or straight"; "Vodka that would compliment two pearl onions on a pick"; "Caviar vodka";"Lovely balance, approachable yet complex"; "Will make a brilliant Flame of Love"

KAI

Style and Origin	New World; Vietnam
Character Score	3
Cost	$$$
Materials & Techniques	Yellow Blossom rice; deionized water; triple distilled—pot initially then twice column distilled; carbon filtration
Tasting Notes	
Nose	Floral notes of orange blossom and honeysuckle; sweet cream; rice pudding; coriander; vanilla; caramel
Palate	Butterscotch sweetness; vanilla; sweet orange; ripe honeydew melon; a touch of mint and black pepper spice
Mouth Feel	Light to medium with a mildly creamy texture and good acidity
Finish	Medium with the slightest touch of spice
Expert Notes	"Mixable, maybe with just chilled soda water, tonic, or ginger beer and a slice of orange"; "Drink it neat paired with spicy Asian-inspired fair"; "I like this for the Hurlyburly"; "Very easy to drink, fun, and approachable"

NEMIROFF ORIGINAL

Style and Origin	Old World; Ukraine
Character Score	7
Cost	$$
Materials & Techniques	Wheat with honey and caraway seeds; purified artesian well water; five-times distilled; filtration proprietary
Tasting Notes	
Nose	Eucalyptus and honeysuckle rounded with ripe honeydew melon, lemon oil, bread-dough grains, vanilla, and white pepper
Palate	Cereal graininess; vanilla richness with caramel and butterscotch; accented by hints of cinnamon, caraway, and white pepper spice
Mouth Feel	Soft, lush, and velvety with very nice acidity
Finish	Medium to long; pleasant; warm with lingering spice
Expert Notes	"Served neat, on the rocks or straight up"; "White Spider—vodka and crème de menthe"; "Dirty Martini"; "Nice in a Gypsy Queen or a Ruby—anything with herbal modifiers"; "Complex yet elegant"

VERMONT GOLD

Style and Origin	New World; Quechee, Vermont, USA
Character Score	5.5
Cost	$$$
Materials & Techniques	100 percent maple sap; spring water; triple distilled in batches with fractioning columns; lightly charcoal filtered
Tasting Notes	
Nose	Prominent toffee and butterscotch with vanilla, cinnamon, caramel, baked apple, and orange marmalade; tropical fruits— mango and ripe pineapple
Palate	Clover honey, toffee, caramel apple, cinnamon, and nutmeg with a touch of black pepper and candied orange rind
Mouth Feel	Medium weight, soft; slightly silky with bright acidity
Finish	Medium and warm with a memory of bittersweet chocolate
Expert Notes	"Lovely on its own or with a pristine cube of ice"; "Desert style cocktails— Espresso Martini, Mudslide, and Black Boot"; "Great with ginger, perfect for the Cucumber Cobbler"; "Complex with a soft side"

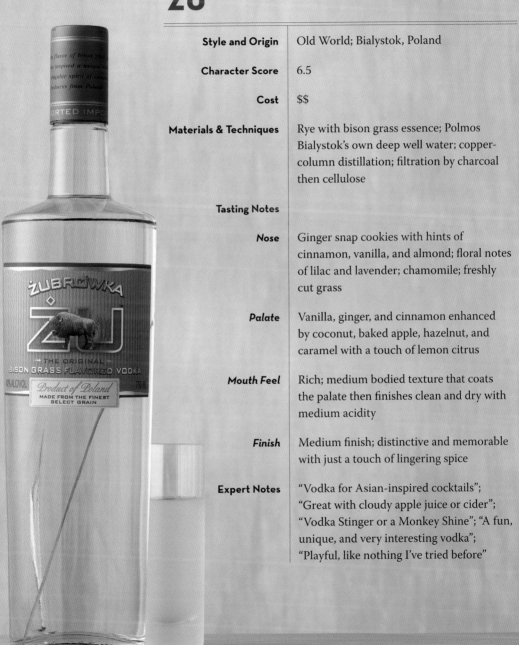

ZU

Style and Origin	Old World; Bialystok, Poland
Character Score	6.5
Cost	$$
Materials & Techniques	Rye with bison grass essence; Polmos Bialystok's own deep well water; copper-column distillation; filtration by charcoal then cellulose
Tasting Notes	
Nose	Ginger snap cookies with hints of cinnamon, vanilla, and almond; floral notes of lilac and lavender; chamomile; freshly cut grass
Palate	Vanilla, ginger, and cinnamon enhanced by coconut, baked apple, hazelnut, and caramel with a touch of lemon citrus
Mouth Feel	Rich; medium bodied texture that coats the palate then finishes clean and dry with medium acidity
Finish	Medium finish; distinctive and memorable with just a touch of lingering spice
Expert Notes	"Vodka for Asian-inspired cocktails"; "Great with cloudy apple juice or cider"; "Vodka Stinger or a Monkey Shine"; "A fun, unique, and very interesting vodka"; "Playful, like nothing I've tried before"

RESOURCES

Industry Websites

AKA Wine Geek
Steve Olson
www.akawinegeek.com

The Barkeeper
Brian Rea's monthly cocktail newsletter
www.thebarkeeper.com

Beverage Alcohol Resource
Comprehensive spirit and mixology education
www.beveragealcoholresource.com

Cocktail
Great database for cocktail recipes, histories,
and antidotes
www.cocktail.com

CocktailDB
Extensive cocktail recipe library
www.cocktaildb.com

Diffords Guide
Extremely concise forum on cocktails
and bartending
www.diffordsguide.com

Drink Boy
Robert Hess's cocktail site
www.drinkboy.com

Gaz Regan
Gaz Regan's spirits and cocktail newsletter
www.gazregan.com

International Spirits Challenge
www.internationalspiritschallenge.com

International Wine and Spirits Competition
www.iwsc.net

Kathy Casey Food Studios—Liquid Kitchen
Exciting and fresh cocktails inspired
by the kitchen
www.kathycasey.com

King Cocktail
Dale DeGroff's website
www.kingcocktail.com

Liquor.com
Articles and recipes by world-renowned
bartenders and experts
www.liquor.com

Market Fresh Mixology
Cocktail books and more
www.marketfreshmixology.com

Mixellany
Jared Brown and Anistatia Miller's
informative and entertaining website
www.mixellany.com

The Modern Mixologist
Tony Abou-Ganim's website
www.themodernmixologist.com

**The Museum of the
American Cocktail**
MOTAC is a nonprofit organization
that celebrates this true American
cultural icon
www.museumoftheamericancocktail.org

**San Francisco World Spirits
Competition**
www.sfspiritscomp.com

Small Screen Network
Web-based shows focused on cocktails
www.smallscreennetwork.com

Spirit Journal
F. Paul Pacult's spirits newsletter
www.spiritjournal.com

Tales of the Cocktail
One of the biggest gatherings of all things cocktail
held annually in New Orleans
www.talesofthecocktail.com

The Tony Abou-Ganim Southwest Spirit Summit"
http://www.facebook.com/tagsummit

Ultimate Spirits Challenge
www.ultimate-beverage.com

United States Bartenders Guild
USBG is an organization of beverage professionals dedicated to the refinement of the craft
www.usbg.org

Webtender
Online drink forum pertaining to the bar
www.webtender.com

Products, Ingredients, Bar Tools, and Miscellaneous Supplies

Angostura Bitters
Aromatic and Orange Bitters
www.angosturabitters.com

Bar Products
Large resource for bar tools
www.barproducts.com

Buffalo Trace
Peychaud's and Regan's orange bitters
www.buffalotracedistillery.com/giftshop.asp

California Caviar
Distributors and producers of sustainable caviar
www.californiacaviar.com

Cocktail Kingdom
Reproduction vintage cocktail books, barware, bitters, and Moscow Mule copper mugs
www.cocktailkingdom.com

Dirty Sue
Olive juice for dirty martinis
www.dirtysue.com

Earthy Delights
Dried hibiscus flowers
www.earthy.com

Employees Only Brands
All natural grenadine syrup and lime cordial
www.employeesonlybrands.com

Fee Brother's
Bitter's, cordial syrups, and cocktail mixes
www.feebrothers.com

Fever Tree
Ginger beer, tonic water, soda water, and bitter lemon
www.fever-tree.com

Filthy Foods
Cocktail onions, stuffed olives, and cherries
www.filthyfood.com

Mister Mojito
Large selection of assorted hardwood muddlers
www.mistermojito.com

The Modern Mixologist
Professional bar tools designed by Tony Abou-Ganim
www.themodernmixologist.com

Monin
Cocktail syrups
www.monin.com

Perfect Purée of Napa Valley
Fresh-fruit frozen purées
www.perfectpuree.com

Petrossian Caviar
Caviar, foie gras, pâté, and smoked fish
www.petrossian.com

Sonoma Syrups Co.
Fruit and spice syrups
www.sonomasyrup.com

BIBLIOGRAPHY

Books

Abou-Ganim, Tony. *Modern Mixologist: Contemporary Classic Cocktails.* Surrey Books. Chicago: 2009

Albert, Bridget. *Market-Fresh Mixology: Cocktails for Every Season.* Surrey Books. Chicago: 2008

Begg, Desmond. *The Vodka Companion.* Running Press. Philadelphia: 1998

Blue, Anthony Dias. *The Complete Book of Mixed Drinks.* Harper Perennial. New York: 1993

Broom, Dave. *The Connoisseur's Book of Spirits & Cocktails.* Carlton Books. London: 1998

Calabrese, Salvatore. *Classic Cocktails.* Sterling. New York: 1997

Carey, Richard Adams. *The Philosopher Fish: Sturgeon, Caviar, and the Geography of Desire.* Counterpoint. New York: 2005

Cecchini, Toby. *Cosmopolitan: A Bartender's Life.* Broadway Books. New York: 2003

Conrad, Barnaby, III. *The Martini.* Chronicle Books. San Francisco: 1995

Craddock, Harry. *The Savoy Cocktail Book.* Lowe & Brydone LTD. London: 1930

DeGroff, Dale. *The Craft of the Cocktail.* Clarkson Potter. New York: 2002

DeGroff, Dale. *The Essential Cocktail.* Clarkson Potter. New York: 2008

Duffy, Patrick Gavin. *The Official Mixer's Manual.* Doubleday & Co. New York: 1956

Edmunds, Lowell. *Martini, Straight Up: The Classic American Cocktail.* Johns Hopkins University Press. Baltimore: 1998

Embury, David A. *The Fine Art of Mixing Drinks.* Doubleday & Co. New York: 1948

Emmons, Bob. *The Book of Gins and Vodkas.* Open Court Publishing. Chicago: 2000

Faith, Nicholas, and Ian Wisniewski. *Classic Vodka.* Trafalgar Square Publishing. London: 1997

Fletcher, Nichola. *Caviar: A Global History.* Reaktion Books. London: 2010

Grimes, William. *Straight Up or On The Rocks: A Cultural History of American Drink.* Simon & Schuster. New York: 1993

Haigh, Ted. *Vintage Spirits & Forgotten Cocktails.* Quarry Books. Beverly, MA: 2004, 2009

Himelstein, Linda. *The King of Vodka: The Story of Pyotr Smirnov and the Upheaval of an Empire.* HarperCollins. New York: 2009

Kosmas, Jason, and Dushan Zaric. *Speakeasy: Classic Cocktails Reimagined, from New York's Employees Only Bar.* Ten Speed Press. Berkeley, CA: 2010

MacElhone, Harry. *Harry's ABC of Mixing Cocktails.* Souvenir Press. London: 1919, 1986

Meehan, Jim. *The PDT Cocktail Book: The Complete Bartender's Guide from the Celebrated Speakeasy.* Sterling Epicure. New York: 2011

Miller, Anistatia, Jared Brown, and Miranda Dickson. *My Absolut: What the Bartender Saw.* Published by the Absolut Company. London: 2010

Miller, Anistatia, and Jared Brown. *Spirituous Journey: A History of Drink, Book Two.* Mixellany. London: 2009

Pacult, Paul F. *Kindred Spirits.* Hyperion. New York: 1997

Pokhlebkin, William. *A History of Vodka.* Verso. London: 1992

Regan, Gary, and Mardee Haidin Regan. *The Martini Companion.* Running Press. Philadelphia: 1997

Regan, Gary. *The Joy of Mixology.* Clarkson Potter. New York: 2003

Saffron, Inga. *Caviar: The Strange History and Uncertain Future of the World's Most Coveted Delicacy.* Broadway Books. New York: 2002

Waggoner, Susan, and Robert Markel. *Make Mine Vodka: 250 Classic Cocktails and Cutting-Edge Infusions.* Stewart, Tabori & Chang. New York: 2006

Walton, Stuart. *Vodka Classified: A Vodka Lover's Companion.* Anova Books Co. London: 2009

Wilson, Jason. *Boozehound: On the Trail of the Rare, the Obscure, and the Overrated in Spirits.* Ten Speed Press. Berkeley, CA: 2010

Wondrich, David. *Esquire Drinks: An Opinionated & Irreverent Guide to Drinking.* Hearst Books. New York: 2002

Websites

www.earthbar.org/vodkahistory.htm

www.freerepublic.com/focus/f-chat/1617176/posts

http://chemistry.about.com/od/famouschemists/p/mendeleevbio.htm

http://absolutad.com/absolut_about/history/story

http://homedistiller.org/ukraine.htm

www.nytimes.com/1995/02/19/nyregion/smirnoff-white-whiskey-no-smell-no-taste.html?pagewanted=all&src=pm

http://online.wsj.com/article/SB118133789715129488.html

www.actuallyalcohol.com/smirnoff-vodka-and-the-american-resurrection

www.diffordsguide.com/cocktail-results.jsp?id=1366

http://liquor.com/liquor/behind-the-drink-the-moscow-mule

http://hackbrew.blogspot.com/2006/09/conversion-of-starch-to-sugar.html

www.physics.uq.edu.au/people/nieminen/vodkamore.html

http://news.bbc.co.uk/2/hi/europe/5333756.stm

http://partydrinks.blogspot.com/2007/03/espresso-martini.html

www.intoxicology.co.uk/recipes/Classics/espresso-martini

www.worldsbestbars.com/cocktails-and-more/Vodka-Espresso.htm

www.ginvodka.org/

www.tastings.com/spirits/vodka.html

www.nytimes.com/2007/06/19/world/europe/19iht-vodka.4.6216260.html?_r=1

www.euractiv.com/food/calling-vodka-vodka-eu-agrees-new-spirit-drinks-regulation/article-169194

www.ginvodka.org/history/vodkaHistory.asp

www.food.com/recipe/el-macua-cocktail-315040

http://projects.washingtonpost.com/recipes/2009/04/08/el-macua/

http://www.campari.com/

www.accidentalhedonist.com/index.php/2005/08/31/history_of_campari

www.esquire.com/drinks/sgroppino-drink-recipe#ixzz1qkgDmK3

http://caviarlover.com/content.aspx?cntid=14

www.sterlingcaviar.com/sterlingcaviar/caviarhistory.asp

www.docstoc.com/docs/2257981/Such-Stuff-as-Dreams-are-Made-OnThe-Story-of-Caviar-from

www.reluctantgourmet.com/caviar.htm

www.fws.gov/le/publicbulletin/PBAvailabilityofFactsheetonBuyingImportedCaviar.pdf

http://rockhudsonblog.com/?tag=elizabeth-taylor&paged=2

www.theramblingepicure.com/archives/5445

www.alexstubb.com/artikkelit/bw_vodka.pdf

ACKNOWLEDGMENTS

So many thanks are due, so we'll just get on with it. Mary and I send our sincere thanks and deep appreciation to…

Alexandre Petrossian and Petrossian Caviar for such generosity of time and product.

All vodka brands, brand ambassadors, distillers, and marketing and PR folks who helped deliver their vital statistics—Angus, you too!

Ally, our culinary genius, for creating a fabulous quick-and-easy blini recipe.

Carol, Christina, and Elaine for organizing and pouring all those vodkas—a very smooth tasting it was.

Claire and Peter—your lipstick and libations we could not have done without.

Dale, Steve, and Bridget for lending so many hours and your collective immeasurable expertise.

Dave (Sweets), Janet, Big A, and Katie for years of support unending.

Dave G. for sharing so much of your limitless industry savvy.

Eric Schmidt and Donna Hood Crecca for their wealth of vodka knowledge.

Everyone at Agate Publishing and Tim Turner Studios—you all rock!

Everyone at the Brass Rail Bar in Port Huron, Michigan.

Floyd for your remarkable Sgroppinos and ongoing inspiration in life and word.

Fort—loyal and ever present, we couldn't have made it without you buddy.

Katie—everyone needs a cheerleader, and you have been unwavering, really priceless.

Miranda Dickson for showing me Poland and turning me on to the beauty of the pork knuckle

Peter and John—your weather reports from across the pond always warm the soul.

Polina Steier at Caviar Affair Magazine for helping source some fabulous products.

Romy and Denise for taking extraordinary care of H.R.H. Vixen, again and again.

Kathy Casey for sharing your innovative H2O knowledge.

San Francisco World Spirits Competition for the reliable delivery of "extreme palate sports."

Simon Ford for the first-hand view of vodka in Sweden and Poland.

SL for delivering Himelstein's inspired look into the realm of Smirnov.

Steelite International and Cardinal Glassware companies for the amazing array from which to choose.

Team EBPMA and ABSPC—finally finished; Sallie, hope you are again inspired!

And, of course, Team *Modern Mixologist*: Andrea, Carol, and Bradley for always keeping the wheels greased and turning.

INDEX